Theological and Ethical Perspectives on Climate Engineering

Religious Ethics and Environmental Challenges
Series Editors: Sarah Fredericks and Kevin J. O'Brien

Religion shapes human responses to twenty-first century environmental challenges—discouraging some adherents from accepting scientific evidence, encouraging others to make sacrifices to preserve ecosystems, and leading still others to develop new spiritual traditions. This interdisciplinary series explores the ways diverse religious communities can, should, and do respond to contemporary environmental challenges. Many of the works will be explicitly ethical, dealing with normative commitments, applied ethics, or ethical theory; others will be theological or philosophical; still others may be social scientific descriptions. Since readers of the series will come from diverse academic contexts, all works will be explicit about methodology, enabling conversation across disciplines. We are particularly interested in works that 1) bring together distinct branches of scholarship to address practical or theoretical issues that cannot be addressed by one alone (e.g., linking healthcare ethics and environmental ethics or comparing religious traditions); 2) explore underresearched religious communities, subcommunities, and traditions; or 3) investigate commonly studied religions in a novel way. We welcome monographs, edited volumes, and exemplary revised dissertations that take one of these approaches. While not all works in the series need to be normative or contemporary, all will help readers advance conversations about the ways religion aids or hinders responses to contemporary environmental challenges.

Recent Titles in This Series

Theological and Ethical Perspectives on Climate Change: Calming the Storm,
Edited by Forrest Clingerman and Kevin J. O'Brien

Theological and Ethical Perspectives on Climate Engineering

Calming the Storm

Edited by Forrest Clingerman
and Kevin J. O'Brien

LEXINGTON BOOKS
Lanham • Boulder • New York • London

Published by Lexington Books
An imprint of The Rowman & Littlefield Publishing Group, Inc.
4501 Forbes Boulevard, Suite 200, Lanham, Maryland 20706
www.rowman.com

Unit A, Whitacre Mews, 26-34 Stannary Street, London SE11 4AB

British Library Cataloguing in Publication Information Available

Library of Congress Cataloging-in-Publication Data Available

ISBN 978-1-4985-2358-5 (cloth : alk. paper)
ISBN 978-1-4985-2359-2 (electronic)

∞™ The paper used in this publication meets the minimum requirements of American National Standard for Information Sciences Permanence of Paper for Printed Library Materials, ANSI/NISO Z39.48-1992.

Printed in the United States of America

Contents

Acknowledgments

This book has been an ongoing and expanding conversation, and we are grateful to all who have joined in.

It began as a conversation between the editors, but quickly expanded when we were invited to lead a track at the 2015 *Seizing an Alternative* Conference in Claremont, California. We are grateful to John Cobb for that invitation and for making that conference such a fruitful setting for conversation, and to Marit Trelstad for connecting us with Dr. Cobb.

The conversation at that conference was incredibly fruitful because of the active participation of many voices. Some are included here, and we are grateful to Thomas Bruhn, Dane Scott, Marit Trelstad, Laura Hartman, and Sarah Fredericks for their writing, commenting, and discussion of these ideas. Others participated in those conversations but chose not to publish here, and we particularly thank Jim Stewart of PODenergy, Sr. Jo Ann Showalter of the Sisters of Providence, and Toby Svoboda of Fairfield University. We also thank Willis Jenkins and Stefan Skrimshire for contributing papers that advanced that discussion and enrich this book.

Ohio Northern University and Pacific Lutheran University gave us each a context from which to work, and we thank our colleagues and students for ongoing conversation and challenges. In particular, we are thankful for the insights of Raymond Person, Mark H. Dixon, students who have taken the ONU class on Christianity and the Natural World, and students who have taken the PLU class on Christian Ecological Ethics.

Even though the conversation around religion and climate engineering is new, discussions of theology, ethics, and environment are not. Many of the relationships that allowed this project to move forward were built thanks due to numerous colleagues and mentors—too many to list—from the American Academy of Religion's Religion and Ecology group; the Forum on Religion and Ecology; the International Society for the Study of Religion, Nature, and Culture, and the European Forum for the Study of Religion and the Environment. We have also benefitted from relationships with colleagues outside religion and theology thanks to the Association for Environmental Studies and Sciences and the Institute for Advanced Sustainability Studies in Potsdam, Germany. The scholars at IASS-Potsdam deserve a particular note of thanks for their role in promoting the discussion of religion and climate engineering.

We are grateful to Sarah Craig and the rest of the staff at Lexington Press for their diligent work, encouragement, and advice in publishing this collection. We also thank Dawn Stover for her excellent editorial guidance on the first chapter, which previously appeared in the *Bulletin of the Atomic Scientists*.

In a book on climate engineering—a book that raises questions about how to envision the human future on this planet—it should come as no surprise that we owe a large debt to friends and family. These are the people for whom we most immediately desire a just and benevolent future, and who have allowed us the time and space to reflect on these issues. We thank them for helping us to remember what we are working for. Particular thanks go to Gail, Asa, and Sabina Clingerman and to Mary O'Brien.

Introduction

Forrest Clingerman & Kevin J. O'Brien

> [Jesus] got into a boat with his disciples, and he said to them, 'Let us go across to the other side of the lake.' So they put out, and while they were sailing he fell asleep. A windstorm swept down on the lake, and the boat was filling up with water, and they were in danger. They went to him and woke him up, shouting 'Master, Master, we are perishing!' And he woke up and rebuked the wind and the raging waves; they ceased, and there was calm. He said to them, 'Where is your faith?' They were afraid and amazed and said to one another, 'Who then is this, that he commands even the winds and the water, and they obey him?'
>
> —Luke 8:22–25 (NRSV)

Our planet is a continually changing place, to be sure. New species evolve and drive others to extinction. Continents move, mountains form, rivers find new course through soil and rock. Such constant changes make the relative stability of the recent past—the last ten thousand years or so—even more amazing. Throughout the Holocene Epoch, the climate of the Earth has been stable, which allowed the human species to thrive, continually expanding its range and influence. This stability is now under threat, though, undermined by the very species that has been one of the Holocene's greatest success stories. Human activity has grown to such an extent that it is fundamentally altering the atmosphere's chemical balance, which in turn is changing the climate. This has unpredictable and severe consequences for the weather that shapes every ecosystem, every species, and every human life.

Climate change is a new meeting point between the complexity of planet Earth and the complexity of human behavior. Given its moral, social, and scientific complexity, climate change is a paradigmatic example of a "wicked problem," or "virtually intractable matters characterized by uncertainty over consequences, diverse and multiple engaged interests, conflicting knowledge claims, and high stakes."[1] The resulting challenge is profoundly difficult to understand, much less to address in an organized and sensible manner. Yet address it we must.

In his important book on the ethics of climate change, Stephen Gardiner writes, "[c]limate change is a perfect moral storm. . . . [It is] an event constituted by an unusual convergence of independently harmful factors where this convergence is likely to result in substantial, and possibly catastrophic, negative outcomes."[2] The image of "a perfect moral storm," emphasizes the conjunction of multiple, contentious challenges, complex on their own terms and exponentially more so in their combination. Indeed, we might say that climate change is a wicked convergence of wicked problems. Gardiner focuses on three moral challenges that come together in the changing climate: first is the challenge that poorer nations and poorer peoples are disproportionately harmed by climate change, but it is disproportionately caused by the rich. This calls for a thoughtful response to global inequality. Second, climate change also creates an intergenerational and interspecies challenge, because it reshapes the planet in irreversible ways. This requires a sensible account of contemporary people's moral responsibility to future generations, other species, and ecosystems. Third, climate change also requires global coordination, creating a theoretical challenge for establishing moral agreement across cultural boundaries. This calls for a thoughtful response to human diversity.

Of course, as Gardiner would acknowledge, the complexities of climate change are not merely moral. This perfect storm also connects to a host of other questions, including the political—what governance structures should and could be accountable to the entire planet?—economic—how should "development" and "growth" be understood in a world of atmospheric limits?—scientific—what risks from a changing climate can reliably be predicted, and what is beyond the scope of current human knowledge?—and psychological—how can people understand a problem this global and complex, and what does it do to their view of the world and one another? Each question, each approach to thinking about climate change, contributes to the complexity and challenge of how to understand and respond to the problem.

In the face of such a storm this perfect, a problem this wicked, what are we to do?

Specialists from science, policy, economics, and ethics proffer myriad answers. But none is self-evident, none is uncontroversial. Mike Hulme has suggested that our disagreements about climate change do not come about because one side is making an error of fact; rather, the disagreement emerges from differences in worldviews and perspectives on such things as science, spirituality, economics, and politics.[3] This means we are unlikely to uncover a simple response that is amenable to all. As a species, we do not know what to do about climate change.

This book will not calm the perfect storm of climate change, but rather takes its subtitle from the quote at the start of this introduction, which is one example of the ways religions offer tools to discuss the complexities of the issue. In the story from the Gospel of Luke, Jesus miraculously soothes a tempestuous sea. The text is almost two thousand years old; its human author would have understood neither globally scaled, anthropogenic climatic change nor the technologies human beings now have at our disposal to cause and, perhaps, solve such a problem. However, texts like this are reinterpreted across centuries, made relevant in ways the original writer could not have intended. For example, an activist seeking to slow or halt the human behaviors that cause climate change might note that only Jesus, God incarnate, could control the Sea of Galilee. This could inform a cautious view of climate change as a sign of hubris, warning against any further attempt to play God by reshaping the oceans and the winds. On the other hand, an advocate of active and adaptive responses to climate change might urge people to follow Jesus's example, to cease cowering in the metaphorical hull and to use our technological prowess to "rebuke the winds and the raging waves."

A third interpretation of this text may be less divisive: faced with a rising storm, fearing impending danger, many people will seek guidance and direction from religious leaders. The disciples in the boat sought instruction from Jesus. Many who are facing the literal and metaphorical storms of climate change similarly seek instruction from ancient religious traditions and the contemporary communities that teach about them. Religion can help to frame, structure, and guide arguments about how to respond to the perfect storm of climate change. Traditions of morality and meaning that have been passed down for generations and millennia can help us to understand, discuss, and make thoughtful decisions about contemporary challenges.

This book brings religious ethicists, theologians, and philosophers together to reflect on a particular set of responses to climate change called "climate engineering" or "geoengineering"—that is, proposals to engineer the global atmosphere in intentional, large-scale ways. Together, the following essays reflect on a common question: In the face of an atmosphere unintentionally altered by human behaviors, is it appropriate to now develop technologies that will intentionally alter the climate even further?

There is no easy answer—again, as a species, we do not know what to do about climate change. *Theological and Ethical Perspectives on Climate Engineering* will not offer a definitive "yes" or "no" to the question of climate engineering. It will, instead, explore how the study of religion—most particularly of Christianity—can contribute to framing and asking the question. In the remainder of this introduction, we will introduce the debate over climate engineering, begin to develop the argument that religion belongs in that debate, and outline the theological and ethical goals of the essays that follow.

THE DEBATE OVER CLIMATE ENGINEERING

At the Earth Summit in 1992, the global community first formally recognized that continued greenhouse gas emissions lead to consequences that are potentially catastrophic for many humans, nonhuman creatures, and ecosystems. At least since then, people have been seeking solutions. But what that solution might be—or even who should have the power to propose or carry out a solution—remains a topic of vigorous debate.

Solutions to climate change are typically discussed in two categories: mitigation and adaptation. The mitigation of climate change seeks to reduce the emissions of greenhouse gases. This can include actions at many scales: from household changes (riding a bike or becoming a vegetarian) to large-scale, societal actions (constructing alternative energy infrastructure, incentivizing energy conservation, or investing in new forms of transportation). In a few cases, such large-scale actions attempt to balance ongoing emissions by removing greenhouse gases (almost always CO_2) from the atmosphere: so-called "carbon credits" are an example of this. Mitigation seeks to reduce (or better yet, eliminate) the causes of anthropogenic climate change and so to avoid possible tipping points that will destabilize the climate. Because of this, mitigation is usually seen as the most preferable response, though it is also the one that requires the greatest potential sacrifices and changes to societal structures.

Proposals for adaptation, in contrast, accept that some of the effects of climate change are inevitable, and propose to adjust human societies to this new world. This includes such projects as building seawalls and levees to deal with rising oceans, addressing new public health issues (e.g., increased ranges for diseases like malaria or heat waves), and climate-related migration of human and nonhuman populations. Adaptation also includes helping farmers to learn to grow crops appropriate to a changing climate or, potentially, moving farmers to new growing areas. Because of the social pressures that will be felt by climate change, adaptation is often seen as unfortunately necessary but less desirable than mitigation.

International commitments to both mitigation and adaptation were made in Paris in December 2015. The agreement, signed by 196 nations, committed to limiting global temperature increases to "well below 2°C above pre-industrial levels," a mitigation plan which involved reducing greenhouse gas emissions. In addition, the agreement acknowledged that the wealthiest nations should continue helping poorer nations to adjust to the world's changing climate, affirming the vital place of adaptation.[4] However, critics and skeptics worry that the Paris agreement is inadequate, that 2°C is too big a climatic shift to expect realistic adaptation in some parts of the world, and that an international commitment with no substantial structure of international law behind it cannot address the perfect storm of climate change.

In the face of such fears about the international political process of mitigation and adaptation, some climate scientists, earth systems engineers, and policymakers have begun developing a third option to combat runaway climate change. Climate engineering, sometimes called geoengineering, is defined by the British Royal Society as "the deliberate large-scale manipulation of the planetary environment to counteract anthropogenic climate change."[5] Climate engineering is, first of all, a *deliberate* attempt to control aspects of the Earth system. People have been manipulating the planetary environment for centuries, as anthropogenic climate change demonstrates. But these changes have been largely unknowing—most of the history of fossil fuel burning took place before climate change was understood—or incidental—even those of us who now burn fossil fuels fully aware of their impact do not do so with the intention of changing the climate. Climate engineering would seek to redress these uninformed and unintentional changes with carefully informed and intentional changes, directly and purposefully reshaping the planet's climate patterns using some of the same mechanisms that underlie the crisis.

Climate engineering is also *large-scale*. Humans have sought to change their weather throughout the history of civilization—cloud seeding, prayers during drought, and seawalls are all attempts to control the weather or its impacts. By contrast, climate engineering seeks to match the scale of global climate change with a global solution, altering the climate (an ongoing, generalized set of phenomena for a region or the entire planet) rather than localized weather events. Local communities may seek to reduce their footprint by painting roofs white to reflect more light or planting a few trees to absorb more CO_2, but climate engineering is fundamentally different because of its magnitude. It seeks to change the composition or the workings of the entire planetary atmosphere.

Finally, climate engineering involves the intentional *manipulation* of the planetary environment, suggesting no simple restoration of previous patterns or removal of human influence, but an ongoing technological effort to influence earth's atmosphere. This technological aspect of climate engineering is what differentiates it from certain mitigation efforts. Large-scale reforestation efforts are a typical example of mitigation, because trees will absorb carbon and thereby mitigate human emissions by restoring an ecosystem. Climate engineering, by contrast, might include the planting of artificial or genetically manipulated trees designed to absorb more CO_2. Climate engineers do not seek to bring about a newly "natural" atmospheric balance; instead, they want to use the tools of twenty-first-century technology to harness, mimic, or replace natural systems, to take direct control over the climate. Along these lines, the National Research Council of the Academy of Sciences suggests that such proposals should be understood as "climate intervention."[6]

This definition of climate engineering includes a confusing array of strategies, because there are multiple ways to imagine deliberately manipulating the planet's atmosphere on a large scale. Examples of some proposals help to show this diversity: One of the more radical is to launch giant mirrors into space, reflecting more of the sun's light away before it reaches the earth.[7] Another proposal calls for huge quantities of iron to be dumped into the ocean at regular intervals, inspiring blooms of algae that would extract carbon dioxide (CO_2) from the atmosphere.[8] A third seeks to build enormous farms full of artificial trees that would not only pull CO_2 from the air but also sequester it permanently underground.[9] Another involves pumping sulfur dioxide particles into the atmosphere—in essence, mimicking a volcanic eruption—to change the way our planet reflects light from the sun and reduce the amount of solar radiation that enters the lower atmosphere.[10] These are just four examples among many, but they reveal just how different climate engineering proposals can be. Some use already existing technology, while others involve new engineering and scientific leaps. Some would require continuous, indefinite intervention, while others are relatively permanent changes to the Earth's systems.

To sort and make sense of the options, scholars of climate engineering tend to distinguish the proposals into two categories: carbon dioxide removal (CDR) and solar radiation management (SRM). CDR seeks to reduce the cause of climate change, removing a significant quantity of CO_2, the most prominent and pernicious greenhouse gas in the atmosphere. Artificial trees and algae blooms are both examples of this approach. CDR is considered slower-acting, more expensive, and less risky than SRM. It also requires planning for unique technical and political issues surrounding the need to sequester the captured CO_2. Some CDR proposals are based upon natural processes (e.g., genetically modifying trees for carbon capture), while others rely entirely on technology—much of which does not yet exist (e.g., creating ambient air scrubbers). Furthermore, CDR approaches are often tied to efforts at climate mitigation, under the assumption that the cost of removing CO_2 will encourage people to reduce their emissions.

SRM, by contrast, does not change the balance of greenhouse gases in the atmosphere, but instead alters the albedo or reflectivity of the planet. Space mirrors and particle clouds are both examples of this approach. To oversimply things a bit, these proposals seek to reduce the amount of thermal radiation available to trap rather than the amount of gas that traps it. Or to use a healing-related metaphor (which Laura Hartmann discusses further in chapter 7 of this volume), SRM focuses on the symptoms rather than the cure. The advantages of SRM are undeniable, but so too are its limitations: as SRM researcher David Keith and his colleagues write, "it is cheap, fast, and imperfect."[11] SRM primarily focuses on technology at hand, and would be fast-acting. Yet because these methods focus on a single variable—global temperature—there are greater risks and larger potential for unexpected outcomes. Furthermore, SRM would not address some of the other effects of higher concentrations of CO_2: coral bleaching, changes in plant growth, and the like.[12]

While advocates of geoengineering generally see it as distinct from other responses to climate change, they also tend to emphasize that it is complementary with them. Climate engineering will be different than mitigation, insofar as climate engineering proposals do not focus on reducing ongoing greenhouse gas emissions. At the same time, geoengineering is like mitigation in that it attempts to avoid the effects of climate change, thereby reducing the need for adaptation. Thus few advocates of CDR or SRM seek to simply continue emissions-intensive human activity into the foreseeable future, burning fossil fuels and continuing economic growth while mitigating the negative impacts with climate engineering technologies. Instead, most advocates of climate engineering hope for a reduction in human emissions, proposing CDR or SRM—or, in all likelihood, a portfolio of approaches that includes both—as a necessary bridge to new technologies or new ways of life that would eventually make climate engineering less urgent.

So, climate engineering is most often envisioned as an intermediate step; it addresses the immediate threats of the climate crisis, "buying time" while mitigation and adaptation efforts are studied and implemented. Some have called it "Plan B," as Dane Scott discusses in chapter 3 of this volume. A worry, however, is that this temporary climate invention will become a permanent feature of the Anthropocene, when ongoing control over Earth systems is unavoidable for human existence. This worry is acute for SRM, since it attends to only certain effects instead of the root cause of climate change. This is the danger of "lock-in": if climate engineering does not address rising CO_2 concentrations, it could be necessary to continue the intervention indefinitely. Stopping SRM unexpectedly or without proper mitigation would lead to rapid and devastating climate effects.

Global climate engineering is not yet a reality; no one is currently managing the planetary environment in a large-scale and deliberate way. There are also few, if any, historical or contemporary parallels to these proposals—the human species has never before united to do anything at the scale or scope of climate intervention. It is thus appropriate that much of the debate about climate engineering is still abstract, seeking to understand and frame the discussion with broad questions: Would it actually be possible to manage solar radiation or remove CO_2 from the atmosphere? What political body or bodies should have the authority to regulate or monitor such efforts? Who should pay for the substantial costs sure to be incurred? What rights might people who may suffer unintended consequences have, and how would those rights be balanced against those of people currently suffering from climate change? Are human beings intellectually and morally capable of managing the entire climate? These theoretical debates are vital, and are already occurring among ethicists, policy analysts, economists, lawyers, and climate scientists.

However, climate engineering research is not entirely abstract, and many researchers understand their work as directly applicable to humanity's present circumstances. Private investors are already exploring options and developing technologies. Applied physicist David Keith has received extensive funding from the U.S. National Science Foundation and from philanthropist Bill Gates to develop a plan for SRM, spraying alumina particles in the upper atmosphere and thereby create a reflective shield in the atmosphere.[13] In 2012, private businessman Russ George conducted a live experiment for a CDR strategy, pouring one hundred tons of iron sulphate into the Pacific Ocean to test and gain publicity for his climate engineering plans.[14]

Policy thinkers and political leaders are also beginning to take these proposals seriously. Climate engineering is becoming a live political issue, calling for careful attention to the interests of individuals, groups, and society at large. One category of political questions relates to distributive justice: how are harms and benefits allocated? Climate engineering will have many impacts, both positive and negative, and political leaders must discern how these can be balanced. A second category of questioning arises in terms of participatory justice: who makes the decisions? Who will decide when climate engineering is a realistic possibility, how it should be implemented, and who is in charge? This is of particular concern given the fact that a single country, corporation, or wealthy individual could potentially undertake climate engineering without broader international approval.[15] Still a third set of questions involves restorative justice: Will climate engineering make up for the past harms caused by climate change?[16] How will less powerful nations and communities be involved in the decision-making process about climate engineering?[17] Should policies around geoengineering assume the need for reparations to countries that have been damaged by past fossil fuel use? Do countries that historically have been the largest emitters of greenhouse causes have the largest role to play in research and implementation?[18] Scholars are only beginning to grapple with these questions, but a group of scientists already drafted five "Oxford Principles" in 2009, focusing on the governance of climate engineering as a common good, informed by public participation, transparently and independently assessed.[19]

We have touched on only a few of the areas of debate and discussion. It should not be surprising that researchers devoted to understanding how the climate might be engineered raise a host of other questions related to ethics and human responsibility: Is there a "moral hazard" created by such activities, such that people might be less inclined to stop damaging the climate once they believe in technology that can repair it? What institutions can monitor climate engineering experiments that occur in the oceans, in space, and otherwise beyond traditional national borders? Could the money currently being spent on such research better be spent developing cleaner energy technologies, or helping people who are already suffering the effects of climate change to adapt?[20] Debates continue, fueling research into climate engineering.

WHY RELIGION BELONGS IN THE DEBATE

Even as the peer-reviewed scientific literature about climate engineering grows, the authors in this volume see something missing: religion belongs in these discussions of climate engineering, and scholars of religion can play a vital role helping to connect the ideas and practices of religious communities with the abstract and the immediate questions raised by deliberate, large-scale manipulations of the planet's atmosphere.

To date, few religious leaders have commented directly on climate engineering, and few political or philosophical discussions of the subject have incorporated awareness of religious communities, traditions, or texts. The few comments that can be found are perfunctory and dismissive. This is a dangerous lacuna, because religion will, inevitably, influence the ways many people think about, feel in response to, and act upon climate engineering proposals. If advocates and critics of climate engineering want to appeal to the broad public sphere, they will need to take religious ideas and religious institutions seriously.

To expand on this claim, the first chapter of this volume notes that many discussions of climate engineering already assume a religious dimension when they ask if large-scale deliberate manipulation of earth's climate is a form of "playing god." In that chapter, we develop four arguments about the importance of religion to discussions of climate engineering: (1) religion is already influential in discussions of climate change, (2) religion belongs in conversation with science and other ways of knowing when facing an issue as complicated as climate engineering, (3) religious language and traditions are already framing discussions of climate engineering, and (4) climate engineering raises theological and ethical questions.

Since that chapter was first published in the *Bulletin of the Atomic Scientists* in 2014,[21] further developments have made it clear that religion is shaping discussions of climate change and should be part of climate engineering discussions. Consider Pope Francis's 2015 encyclical, *Laudato Si'*, which identified climate change as "one of the principle challenges facing humanity in our day." The encyclical—addressed to all people of good will and especially to the over one billion Catholics that arguably constitute the largest religious community in the world—urges Christians to recognize that their relationship to God cannot be healthy unless they are also caring for their human neighbors and the earth itself. In fact, the encyclical states, the climate is a "common good," "belonging to all and meant for all," necessary for human social and individual fulfillment.[22]

Of course, others use religious ideas to justify less concern about climate change. For example, as the papal encyclical was still being drafted, a group of evangelical thinkers called the Cornwall Alliance composed an open letter encouraging Pope Francis to dismiss any reductions in the use of fossil fuels, asserting that dominant understandings of climate science are simply untrue. They end their letter with a quote from Psalm 19—"The heavens declare the glory of God; and the firmament proclaims his handiwork"—and this interpretation: "By using fossil fuels to generate energy to lift billions of God's precious children out of poverty, we liberate from the tomb of the earth the carbon dioxide on which plants and therefore all the rest of life depend."[23] The theology of this document suggests that God would never allow human technology to fundamentally harm the atmosphere, and that God advocates and supports technologies that make economic growth possible. Such faith leads to deep suspicion about mainstream climate science.

Neither Pope Francis nor the Cornwall Alliance mentions climate engineering. However, they effectively demonstrate that religion is a part of the broad debates about climate change, and that even within one religious tradition there is no widespread or simple agreement about how to respond to the perfect moral storm facing humanity.

In our judgment, the Cornwall Alliance makes a theological mistake, assuming that an interpretation of the Bible should trump clear evidence and expert consensus and thereby giving a particular theology authority over science. However, the opposite mistake, which would give authority entirely to science and ignore the insights and contributions of religion, is also dangerous. Pope Francis makes this point well: "If we are truly concerned to develop an ecology capable of remedying the damage we have done, no branch of the sciences and no form of wisdom can be left out."[24]

The authors in *Theological and Ethical Perspectives on Climate Engineering* seek a collaborative approach between science and religion, assuming that climate engineering raises vital scientific and political questions, but also questions of meaning and value that cannot be fully resolved by the natural and social scientists. Thus, a collaborative approach will be a vital aspect of any nuanced and helpful perspective on the topic.

Religion is already an important part of the way many people think, or decide not to think, about climate change. Religion has a role to play alongside other forms of knowing in any environmental or moral discussion. Thus, it is no surprise that religious imagery and narratives are already used in debates about climate engineering. The more deliberate and explicit such use of religion can be in the future, the more productive such debates will be.

THEOLOGY AND ETHICS: THE BOOK IN OUTLINE

This introduction and the first chapter elaborate the book's foundational premise, that religion is relevant and important to discussions of climate engineering. The remainder of the book assumes that claim and moves on to consider its implications. The second chapter, by Thomas Bruhn and his colleagues from the Institute for Advanced Sustainability Studies in Potsdam, Germany, shows a practical implication of religion's relevance, recounting their attempts to bring religion into transdisciplinary conversations at the institute and lifting up some key contributions to the broader debate that emerged from their discussions.

The book then turns to explicitly theological and ethical analysis. Part II focuses on theology, which twelfth-century monk Anselm of Canterbury defined as "faith seeking understanding." This is as good a definition as has been articulated since. While some scholars of religion begin from the social structures or practices of religious people, theologians seek first and foremost to understand human faith—what do people think about the world, about their own actions in it, and about how both are related to ultimate reality? How do people make meaning of their own existence and the communities and systems of which they are a part? How do communities enact their beliefs in ritual, moral, and daily practices? To do theology is to seek to understand human faith and its implications for the world. Many theologians have offered theological assessments of anthropogenic climate change. For instance, a group led by Michael Northcott and Peter M. Scott have undertaken a discussion of how climate change might influence traditional Christian ideas, such as the doctrines of Christology, the Holy Spirit, and the Church.[25] Likewise, to think theologically about climate engineering is to ask how tenets and practices of faith are shaping and could shape decisions about deliberate, large-scale manipulation of the atmosphere.

Theology has a key role in addressing the potential relevance of religion to climate engineering. Theological reflection offers a systematic, constructive, and critical appraisal of how religion acts as a deep-seated, pervasive lens for believers. For example, a theologian might note that one cannot determine whether or not to engineer the climate without delving into the foundational question of how human beings relate to the other living and nonliving systems of earth. By virtue of religious faith, believers hold that there is meaning to life on earth. But how can we understand what that meaning is? And why do different religious communities disagree on how to respond to such meaning and purpose? Theological reflection takes up these questions, evaluates possible claims to meaning, and normatively constructs a richer account for people inside and outside the religious community. In the case of climate engineering, one view might be that God created human beings as the caretakers of the earth who should express our gratitude for

God's goodness by taking care of what God has made. A very different answer would be that all life is fundamentally interconnected and equal, and that human beings live best when we live as one species among many, fully participating in nature's processes. Still another answer would be that human beings are a product of evolution, motivated by biology to protect our own species by ensuring the survival of our genes into the next generation. These three perspectives on the human relation to the world frame the issue of climate engineering in vastly different ways, suggesting that theological reflection needs to be part of the discussion.[26]

In chapter 3, Dane Scott draws on the theological anthropology of Reinhold Niebuhr to argue that climate engineering should only be discussed in light of a careful and sophisticated awareness of human limitation—in Christian language, sin. Chapter 4 offers a different theological perspective, as Marit Trelstad shows that the integrative worldview of the Christian tradition and process theology raise challenging questions for perspective climate engineers. In chapter 5, Forrest Clingerman moves the discussion to a theology of time, arguing that every claim made about climate engineering involves a narrative assuming a certain pace, shedding light on the nature of disagreements about climate engineering. Stefan Skrimshire draws on similar themes in chapter 6, arguing that climate engineering is parallel to other large-scale efforts to resurrect extinct species, and that both must be understood as attempts to rewrite what the Christian tradition has called "the book of nature."

Part III of the book then deals more explicitly with ethics. Turning from theology to ethics is a move from questions of faith and meaning to questions of action and interaction. Ethics, the study of moral ideas and behavior, asks how people and communities relate to other persons, other creatures, and the world. Religious ethics asks questions such as: What do we owe to other people, to other creatures and species, to the Earth itself? Does a belief in a transcendent God increase or decrease believers' commitment to the immanent world around them? What guidance, if any, can ancient texts and traditions offer to uniquely contemporary moral challenges?

Religious ethicists have grappled extensively with the problem of climate change, and developed distinctly different positions. For example, Carol Robb argues in *Wind, Sun, Soil, Spirit* that Christians must respond to climate change as people "rooted in" the Hebrew Scriptures and New Testament, finding biblical resources for their response to this crisis.[27] By contrast, James Nash argues in "The Bible vs. Biodiversity" that Christians must turn to sources other than scripture—which he finds outdated and in conflict with ecological ideas—to respond to twenty-first-century environmental challenges.[28]

To extend religious ethics into the realm of climate engineering is to continue such questions, asking, for example, if the Biblical texts about which Robb and Nash argue can and should be brought to bear on the deliberate, large-scale manipulation of the climate. Beyond this question of moral sources, it is also important to ask how principles of justice and love and care and obedience—moral principles with profoundly religious roots—apply to climate engineering. The ethical questions of climate engineering are legion, and those raised by philosophers, activists, and politicians have been discussed throughout this introduction. Religious ethics notes that theological ideas and communities of faith can and should contribute to those discussions.

In chapter 7, Laura Hartman challenges the dominant framework used to understand climate engineering, asking what it might mean to see manipulations of the climate as efforts at "healing" rather than "repairing" the atmosphere. In chapter 8, Willis Jenkins questions any overarching narrative for the practice, suggesting that the role of ethics is to ironically shock people to a new awareness that transcends any single framework. Sarah Fredericks appeals to more specific religious moral practices in chapter 9, suggesting that moral responses to climate engineering must take religious rituals seriously. Kevin O'Brien expands on that suggestion in chapter 10, arguing that rituals as well as traditions and ideals of Christian repentance should slow and caution discussions of climate engineering.

ENGAGING CONVERSATIONS ON CLIMATE ENGINEERING

To fully bring religious communities, religious ideas, and religious practices into discussions of climate engineering will be the project of years and decades. We hope that *Theological and Ethical Perspectives on Climate Engineering* contributes to that project, but our immediate goal is more modest: we have brought together scholars who study Christian religion and asked them to reflect theologically and ethically on climate engineering. This book's goal is to nurture sophisticated discussions of climate engineering among other scholars, and perhaps indirectly shape a wider debate.

Because of its interest in religion, ethics, and geoengineering, this book intervenes most particularly in two conversations, with two overlapping but distinct audiences. One is the ongoing conversation among researchers about climate engineering, which has to this point paid little attention to religion. The chapters in this book—particularly those by Clingerman, Hartman, Jenkins, Scott, and Skrimshire—address the philosophers, engineers, and scientists currently analyzing climate engineering proposals, arguing that they will benefit in many ways from integrating religion into their transdisciplinary conversations. The other conversation is among Christian communities concerned about the environment, which have to this point paid little attention to the specific topic of climate engineering. The chapters of this book—

particularly those by Bruhn, Fredericks, O'Brien, and Trelstad—propose that Christians have the tools to learn about and engage climate engineering, and should step actively and thoughtfully into the debate. While each chapter might focus on one or the other conversation, we hope that the book as a whole suggests a model for how religion scholars can serve as intermediaries between religious and scientific communities.

Both these conversations—the scholarly conversation about climate engineering and the religious conversation about environmental challenges—are ongoing, global, and complex. Thus, this book's contributions are necessarily partial. Its authors are all from Europe and the United States, academics with at least moderate privilege, power, and security. The book does not include perspectives from other parts of the world, nor from oppressed and marginalized regions or peoples, though several articles aim to be aware of such perspectives. More work remains to be done to assess the place of religion in debates about how climate engineering is related to issues of ecojustice, reparations, and global governance. We recognize this as a sign that the work here is incomplete, and we hope that this project contributes to a dialogue in which other voices will feel empowered to challenge our assumptions and conclusions.

Along similar lines, the chapters of this book are written by scholars who seek to understand religion broadly and Christianity in particular, and our primary vocabulary is taken from religious studies and theology. This focus on Christianity is intentional, even though we agree with Pak-Hang Wong that a truly global ethic around climate engineering will require investigating many traditions, including non-Western religions.[29] Analysis focusing on other religious traditions and commentary from other academic disciplines are vital projects, but beyond the scope of this book.

Theological and Ethical Perspectives on Climate Engineering is, above all, a testament to what can come from good conversations. The essays were first shared when most of the authors gathered in Claremont, California, in the summer of 2015 for an intensive three days of exchange, debate, and analysis.[30] Every author has continued exchanging and refining ideas since, and we are grateful to them for all the ways we have learned and developed as a community. Readers will notice that we do not by any means agree about climate engineering. While all of us raise questions for the most jubilant advocates of climate engineering, some of us began the project with complete skepticism about any proposals to manipulate the atmosphere and others were more open to the possibilities of climate engineering. In the course of the conversations, some have changed our minds, some have not. Our focus throughout has not been on finding clear answers, but rather structuring a productive conversation in which the complexities and importance of bringing religion into the climate engineering debate become clearer.

In the course of our conversations, some ideas came out clearly as common ground, shared principles and questions we brought to the issue of climate engineering. That common ground was drafted into a document by Marit Trelstad and Thomas Bruhn, and is included as the appendix to this book, "Religion and Climate Engineering: Points of Consensus from Claremont."

These points of agreement served as a helpful foundation for conversation, but that conversation was engaging and important because we deeply explored our disagreements and the questions that none of us feel ready to answer. The authors in this book agree that climate engineering is worthy of discussion, that it is neither so monstrous as to be outside of consideration nor so obvious as to be self-evidently a good idea. But we disagree strongly about where to place ourselves between these two extremes. We agree that any discussion of climate engineering needs a clear narrative framework, but we disagree about which frameworks are most promising and useful. We agree that religion has a place in conversations about climate engineering, but we come from different religious and areligious backgrounds, and we understand religious practices and beliefs in very different ways. We agree that no single interest group or academic discipline should control discussions of climate engineering, but we disagree on how broadly conversation can pragmatically be extended. We agree that any response to climate change must consider not just atmospheric forces but also the structural inequalities that shape the contemporary human world, but we disagree on whether and how such power dynamics are determinative of concrete responses to proposals and research on climate engineering.

This book aims to help other scholars engage in further conversations about climate engineering, to extend our conversations so that others can join in. Climate engineering can unfortunately inspire debate in which sides are clearly taken and discussion ended. We hope *Theological and Ethical Perspectives on Climate Engineering* helps instead to nurture a genuine conversation, in which ideas are carefully developed, freely exchanged, and thoughtfully considered.

NOTES

1. Irene Lorenzini, Mavis Jones, and John R. Turnpenny, "Climate Change, Human Genetics, and Postnormality in the UK," *Futures* 39 (2007): 65.
2. Stephen Mark Gardiner, *A Perfect Moral Storm: The Ethical Tragedy of Climate Change* (New York: Oxford University Press, 2011), 22–23.
3. Mike Hulme, *Why We Disagree about Climate Change* (New York: Cambridge, 2009).
4. Conference of the Parties. "The Paris Agreement." December 11 (2015).
5. The British Royal Society, *Geoengineering the Climate: Science, Governance, and Uncertainty* (London: Royal Society Reports, 2009). This book will use the terms geoengineering and climate engineering interchangeably.

6. National Research Council (US), *Climate Intervention: Reflecting Sunlight to Cool Earth* (Washington, DC: National Academies Press, 2015), 1.

7. Roger Angel, "Feasibility of cooling the Earth with a cloud of small spacecraft near the inner Lagrange point (L1)," *PNAS* 103, no. 46 (November 14, 2006): 17184–89; D. J. Lunt, A. Ridgwell, P. J. Valdes, and A. Searle, "'Sunshade World': A Fully Coupled GCM Evaluation of the Climatic Impacts of Geoengineering," *Geophysical Research Letters* 35 (2008): L12710.

8. K. O. Buesseler et al., "Ocean Iron Fertilization—Moving Forward in a Sea of Uncertainty," *Science* 319:5860 (2008).

9. Klaus Lackner, "Washing Carbon Out of the Air," *Scientific American* 302:6 (2010).

10. This proposal is among the most studied, especially after Nobel Laureate Paul J. Crutzen's influential essay, "Albedo Enhancement by Stratospheric Sulfur Injections: A Contribution to Resolve a Policy Dilemma?" *Climatic Change* 77 (2006): 211–9.

11. David W. Keith, Edward Parson, and M. Granger Morgan, "Research on Global Sun Block Needed Now," *Nature* 463 (January 28, 2010): 246.

12. See Alan Robock, Allison Marquardt, Ben Kravitz, and Georgiy Stenchikov, "Benefits, Risks, and Costs of Stratospheric Geoengineering," *Geophysical Research Letters* 36 (2009): L19703; J. Feichter and T. Leisner, "Climate Engineering: A Critical Review of Approaches to Modify the Global Energy Balance," *European Physical Journal Special Topics* 176 (2009): 81–92.

13. David W. Keith, *A Case for Climate Engineering* (Cambridge, MA: MIT Press, 2013). Keith discusses his funding and its implications for his research at http://www.keith.seas.harvard.edu/affiliations.

14. Martin Lukacs, "World's Biggest Geoengineering Experiment 'Violates' UN Rules," October 15 (2012): accessed August 23, 2015, http://www.theguardian.com/environment/2012/oct/15/pacific-iron-fertilisation-geoengineering. For George's own account of his experiment and its purposes, see http://russgeorge.net/.

15. One attempt to wrestle with these challenges is the "Oxford Principles," which argue that climate engineering must be understood and as a public good, with public participation and openness as key for both research and implementation. See Steve Rayner et al., "The Oxford Principles," *Climatic Change* 121 (2013): 499–512.

16. See, for example, David Keith's argument that the rich owe it to the poor to pursue climate engineering projects, in *A Case for Climate Engineering*, chapter 5.

17. Tina Johnson, "A Call for Engagement by Environmental Justice Groups in the Geoengineering Conversation," January 27 (2014), http://dcgeoconsortium.org/2014/01/27/guest-post-tina-johnson-energy-action-coalition-a-call-for-engagement-by-environmental-justice-groups-in-the-geoengineering-conversation/.

18. For extended discussion of the political issues of climate engineering, see especially Wil Burns & Andrew Strauss, eds., *Climate Change Geoengineering: Philosophical Perspectives, Legal Issues, and Governance Frameworks* (New York: Cambridge, 2013), especially chapters 4–12.

19. Steve Rayner et al. "The Oxford Principles."

20. The most influential set questions about climate engineering is Alan Robock, "20 Reasons Why Geoengineering May Be a Bad Idea," *Bulletin of the Atomic Scientists* 64:2 (2008).

21. Forrest Clingerman and Kevin J. O'Brien, "Playing God: Why Religion Belongs in the Climate Engineering Debate." *Bulletin of the Atomic Scientists* 70:3 (2014): 27–37.

22. Pope Francis, *Laudato Si': On Care for Our Common Home* (Vatican City: Encyclical Letter, 2015), § 26 & 156. This encyclical should be seen as a continuation of a long-standing concern over environmental issues. See Jame Schaefer, "Solidarity, Subsidiarity, and Preference for the Poor: Extending Catholic Social Teaching in Response to the Climate Crisis," in *Confronting the Climate Crisis: Catholic Theological Perspectives*, edited by Jame Schaefer (Milwaukee: Marquette University Press, 2011), 389–425; and Christiana Z. Peppard, "Pope Francis and the Fourth Era of the Catholic Church's Engagement with Science," *Bulletin of the Atomic Scientists* 71 (2015): 31–39.

23. The Cornwall Alliance, "An Open Letter to Pope Francis on Climate Change," available at http://www.cornwallalliance.org/anopenlettertopopefrancisonclimatechange/ (accessed Oct. 3, 2015). It is interesting to compare this with William Nordhaus's critique of the encyclical.

Nordhaus does not comment on the theological backdrop, and instead offers an analysis and critique of the document's understanding of market forces. Nordhaus, unlike the Cornwall Alliance, seems to miss the key argument of *Laudato Si*—that one's theological perspective forms the basis for one's interpretation of economics within what Pope Francis calls "integral ecology." See William D. Nordhaus, "The Pope and the Market," *New York Review of Books* 62, no. 15 (October 8, 2015): 26–27.

24. Pope Francis, *Laudato Si': On Care for Our Common Home*.

25. Michael Northcott and Peter Manley Scott, *Systematic Theology and Climate Change: Ecumenical Perspectives* (New York: Routledge, 2014).

26. For an extended discussion of two distinct Christian theologies and their implications for climate engineering, see Forrest Clingerman, "Between Babel and Pelagius: Religion, Theology, and Geoengineering," in *Engineering the Climate: The Ethics of Solar Radiation Management*, ed. Christopher J. Preston (New York: Lexington Books, 2012).

27. Carol S. Robb, *Wind, Sun, Soil, Spirit: Biblical Ethics and Climate Change* (Minneapolis: Fortress Press, 2010).

28. James A. Nash, "The Bible vs. Biodiversity: The Case against Moral Argument from Scripture," *Journal for the Study of Religion, Nature, and Culture* 3:2 (2009).

29. Pak-Hang Wong, "Confucian Environmental Ethics, Climate Engineering, and the 'Playing God' Argument," *Zygon* 50 (2015): 28–41.

30. We thank John Cobb and the Center for Process Studies for inviting us to organize the track on Geoengineering at the 2015 *Seizing an Alternative* conference, which provided an energizing venue for these conversations. We also thank the participants in our conversations who did not end up contributing chapters to the volume, especially Toby Svoboda of Fairfield University and Jim Stewart of PODenergy. Finally, we thank Mark Lawrence and Stefan Schäfer, both of IASS, who have provided several opportunities for discussion and debate on these topics.

BIBLIOGRAPHY

Angel, Roger. "Feasibility of Cooling the Earth with a Cloud of Small Spacecraft Near the Inner Lagrange Point." *PNAS* 103:46 (2006): 17184–89.

The British Royal Society. *Geoengineering the Climate: Science, Governance, and Uncertainty*. London: Royal Society Reports, 2009.

Buesseler, K. O.; S. C. Doney; D. M. Karl; P. W. Boyd; K. Caldeira; F. Chai; K. H. Coale; H. J. De Baar; P. G. Falkowski; K. S. Johnson; R. S. Lampitt; A. F. Michaels; S. W. A. Naqvi; V. Smetacek; S. Takeda; A. J. Watson. "Ocean Iron Fertilization—Moving Forward in a Sea of Uncertainty," *Science* 319:5860 (2007), 162.

Burns, Will C. G., and Andrew L. Strauss. *Climate Change Geoengineering: Philosophical Perspectives, Legal Issues, and Governance Frameworks*. New York: Cambridge University Press, 2013.

Clingerman, Forrest. "Between Babel and Pelagius: Religion, Theology, and Geoengineering." In *Engineering the Climate: The Ethics of Solar Radiation Management*, edited by Christopher J. Preston, 201–19. New York: Lexington Books, 2012.

Clingerman, Forrest, and Kevin J. O'Brien. "Playing God: Why Religion Belongs in the Climate Engineering Debate." *Bulletin of the Atomic Scientists* 70, no. 3 (2014): 27–37.

Conference of the Parties. "The Paris Agreement." December 11 (2015): Accessed December 23, 2015.

The Cornwall Alliance. "An Open Letter to Pope Francis on Climate Change." April 27, 2015. Available at http://www.cornwallalliance.org/anopenlettertopopefrancisonclimatechange/ (accessed October 3, 2015).

Crutzen, Paul J. "Albedo Enhancement by Stratospheric Sulfur Injections: A Contribution to Resolve a Policy Dilemma?" *Climatic Change* 77 (2006): 211–9.

Feichter, J., and T. Leisner. "Climate Engineering: A Critical Review of Approaches to Modify the Global Energy Balance." *European Physical Journal Special Topics* 176 (2009): 81–92.

Gardiner, Stephen Mark. *A Perfect Moral Storm: The Ethical Tragedy of Climate Change*. New York: Oxford University Press, 2011.
Hulme, Mike. *Why We Disagree about Climate Change.* New York: Cambridge, 2009.
IPCC. *Climate Change 2014: Synthesis Report. Contribution of Working Groups I, II and III to the Fifth Assessment Report of the Intergovernmental Panel on Climate Change.* Geneva: IPCC 2014. http://www.ipcc.ch/pdf/assessment-report/ar5/syr/SYR_AR5_FINAL_full.pdf, accessed October 14, 2015.
Johnson, Tina. "A Call for Engagement by Environmental Justice Groups in the Geoengineering Conversation." January 27 (2014). http://dcgeoconsortium.org/2014/01/27/guest-post-tina-johnson-energy-action-coalition-a-call-for-engagement-by-environmental-justice-groups-in-the-geoengineering-conversation/. Accessed October 12, 2015.
Keith, David W. *A Case for Climate Engineering*. Cambridge, MA: MIT Press, 2013.
Keith, David W., E. Parson, and M. G. Morgan. "Research on Global Sun Block Needed Now." *Nature* 463(2010): 245–85.
Lackner, Klaus. "Washing Carbon Out of the Air." *Scientific American* 302:6 (2010), 58–53.
Lorenzini, Irene, Mavis Jones, and John R. Turnpenny. "Climate Change, Human Genetics, and Postnormality in the UK." *Futures* 29 (2007): 65–82.
Lukacs, Martin. "World's Biggest Geoengineering Experiment 'Violates' UN Rules." October 15 (2012): Accessed August 23, 2015. http://www.theguardian.com/environment/2012/oct/15/pacific-iron-fertilisation-geoengineering.
Lunt, D. J., A. Ridgwell, P. J. Valdes, and A. Searl. "'Sunshade World': A Fully Coupled GCM Evaluation of the Climate Impacts of Geoengineering." *Geophysical Research Letters* 35 (2008): L12710.
Nash, James A. "The Bible vs. Biodiversity: The Case against Moral Argument from Scripture." *Journal for the Study of Religion, Nature, and Culture* 3:2 (2009), 213–37.
National Research Council (US). *Climate Intervention: Reflecting Sunlight to Cool Earth.* Washington, DC: National Academies Press, 2015.
Nordhaus, William D. "The Pope and the Market." *New York Review of Books* 62, no. 15 (October 8, 2015): 26–27.
Northcott, Michael, and Peter Manley Scott. *Systematic Theology and Climate Change: Ecumenical Perspectives.* New York: Routledge, 2014.
Peppard, Christiana Z. "Pope Francis and the Fourth Era of the Catholic Church's Engagement with Science." *Bulletin of the Atomic Scientists* 71 (2015): 31–39.
Pope Francis. *Laudato Si': On Care for Our Common Home*. Vatican City: Encyclical Letter, 2015.
Rayner, Steve, Clare Heyward, Tim Kruger, Nick Pidgeon, Catherine Redgwell, Julian Savulescu. "The Oxford Principles." *Climatic Change* 121 (2013): 499–512.
Robb, Carol S. *Wind, Sun, Soil, Spirit: Biblical Ethics and Climate Change* (Minneapolis: Fortress Press, 2010).
Robock, Alan. "20 Reasons Why Geoengineering May Be a Bad Idea." *Bulletin of the Atomic Scientists* 64, no. 2 (2008): 14–18.
Robock, Alan, Allison Marquardt, Ben Kravitz, and Georgiy Stenchikov. "Benefits, Risks, and Costs of Stratospheric Geoengineering." *Geophysical Research Letters* 36 (2009): L19703.
Schaefer, Jame. "Solidarity, Subsidiarity, and Preference for the Poor: Extending Catholic Social Teaching in Response to the Climate Crisis." In *Confronting the Climate Crisis: Catholic Theological Perspectives*, edited by Jame Schaefer (Milwaukee: Marquette University Press, 2011), 389–425.
Wong, Pak-Hang. "Confucian Environmental Ethics, Climate Engineering, and the 'Playing God' Argument." *Zygon* 50 (2015): 28–41.

I

Climate Engineering and Religion

Chapter One

Playing God: Why Religion Belongs in the Climate Engineering Debate[1]

Forrest Clingerman and Kevin J. O'Brien

Arguing against climate legislation in 2009, Congressman Joe Barton of Texas announced: "You can't regulate God. Not even the Democratic majority in the US Congress can regulate God."[2] Barton asserted that only God is in control of the atmosphere's greenhouse gas concentrations and therefore that the effects of climate change are not in the hands of humans. With this claim, Barton represented a significant number of conservative Christians in the United States for whom religion defines the debate about climate change and whose religious beliefs can be the source of opposition to measures aimed at addressing global warming. For good or for ill, religion is part of the discussion of climate change.

Interestingly, however, discussions of the religious implications of climate change have not been meaningfully translated into the public understanding of geoengineering, defined by the British Royal Society as "the deliberate large-scale manipulation of the planetary environment to counteract anthropogenic climate change."[3] Aspiring geoengineers suggest that human beings must actively and intentionally take control of Earth's thermostat in response to the ways industrial emissions have already unintentionally changed it. As this suggestion becomes more widely discussed, it will almost certainly trigger religious responses. Advocates and critics of geoengineering alike will be better served if diverse religious perspectives enter the conversation as soon as possible, both to prevent misinformation coming from religious communities and to draw on the considerable resources religious communities can offer to the discussion.

At some point, religion will play a role in the public perception of geoengineering, through the voices and choices of adherents and spiritual leaders.

If these religious people are in open dialogue with scientists, engineers, and policy makers, they could be valuable contributors to the conversation. But if religious discussions are kept entirely separate from technical policy decisions, neither is likely to be fully productive. This is where the study of religion becomes important: scholars of religious studies and theology can act as mediators between the scientific and faith communities, providing a critical voice in understanding exactly how religion affects the climate debate.

When scientists and political leaders contemplate "playing God" with the climate, they should recognize the ways religion might influence public perceptions of the game. More specifically, scientists discussing geoengineering should pay attention to religion in four ways: as part of how most humans define themselves; as an integral influence on public engagement with science; as a source of conceptual frames; and as a set of traditions for moral discernment.

THE GEOENGINEERING DEBATE

For now, geoengineering is a theoretical idea rather than a practical manifestation, but it has gained momentum since two 2006 publications. The first, by Nobel Prize laureate Paul Crutzen, argued that political inaction creates the need to seriously research geoengineering. Crutzen specifically advocated mimicking the global cooling seen after volcanic eruptions, like that of Mount Pinatubo in 1991. The second 2006 paper, written by Tom Wigley, who directed the Climatic Research Unit at the University of East Anglia before becoming a senior scientist at the National Center for Atmospheric Research in the United States, similarly suggested that mitigation must be combined with geoengineering in response to the severity of anthropogenic climate change. The stature of these authors—both preeminent scientists well known for their work on climate-related issues—caught the attention of the scholarly community. These articles opened the floodgates; geoengineering quickly changed from the subject of fringe speculation to an area of more serious consideration.

Of course, geoengineering proposals are diverse.[4] One form is solar radiation management, reflecting sunlight to compensate for the warming effects of increased greenhouse gases. Solar radiation management is possible with existing technology, but it cannot "restore" the climate and will likely have unintended consequences; as Harvard University climate scientist David Keith and his colleagues explain, solar radiation management is "cheap, fast, and imperfect."[5] In contrast, the slower and more expensive option of carbon dioxide removal concentrates on the underlying cause of climate change. Carbon dioxide removal proposals, which range from filtering ambient air to fertilizing the ocean to create algae blooms, would have more predictable outcomes, but could potentially be far more complicated and expensive. Whatever form it took, geoengineering would require extensive interdisciplinary research.[6] Furthermore, any actual implementation would most likely use several different technological interventions, or what UC San Diego legal scholar David G. Victor calls "cocktail engineering," rather than "silver bullet" approaches.[7]

While scientists continue to model and study the feasibility of geoengineering, many of the most important questions are outside the realm of technical and scientific expertise. Indeed, as *Bulletin of the Atomic Scientists* executive director and publisher Kennette Benedict explains, geoengineering and the problem of climate change are topics for the *Bulletin* because the latter poses an "existential threat" to humanity.[8] Reflection on, and response to, existential threats is not solely the domain of science. Geoengineering is also a political issue, and so there has been an increasing recognition of the need to encourage and examine public engagement with the topic.[9] Ethics is also a concern, as Alan Robock, a climate scientist well known for his critiques of geoengineering, pointed out when he articulated twenty reasons to be concerned about geoengineering.[10]

But something is still missing. Responsible consideration of geoengineering as a scientific, existential, political, and moral issue requires engagement with religion. That makes some people nervous, because it reminds them of religious attacks against science—from Galileo's trial to Congressman Joe Barton's dismissal of climate change. But while religion has had a difficult relationship with science and technology, it also has an incredible influence on the growth of these fields. Religious values are often implicit in public understanding and engagement. Religious narratives provide a vocabulary to orient moral and conceptual understanding of issues. Religious motivations have served to spur innovation. And religious communities are informal conduits for debate and decision-making.

RELIGION DEFINES HUMAN BEINGS AND SOCIETIES

Religion is fundamental to how human beings understand themselves and their place in the world. More than two-thirds of American adults consider themselves at least moderately religious,[11] and 84 percent of the world's population is affiliated religiously.[12] Of course, these numbers represent varying levels of commitment to many different faiths. But despite this diversity, religion is one of the central characteristics of human societies and the human species—we are "Homo religiosus." If advocates and critics of climate engineering want to appeal to the broad public sphere nationally or globally, they will need to communicate with religious people. Further, not only religious beliefs but also religious institutions matter. If advocates and critics of climate engineering want to change cultural practices and political structures, they will need to reckon with the organized faith structures and institutions that have heavily influenced society.

The connection between religion and climate has not gone unnoticed. Religion scholars, particularly of Buddhism and Christianity, have argued that faith traditions have historically offered local, national, and global responses to a changing world and its moral challenges.[13] The converse is also true: Religions can dampen concern for environmental issues broadly, and climate change in particular, with research suggesting that theological affiliation partly determines overall environmental concern.[14] For example, political scientists David Barker and David Bearce have studied how American Christians with a strong "end-times"-oriented theology are less likely to support policies aimed at fighting climate change.[15]

Yet despite this considerable attention to religion and climate change, the role of religious belief in geoengineering debates remains relatively unexplored.[16] Geoengineering raises difficult questions about the place of human beings in relation to the nonhuman world; the potential and limits of human ingenuity; and the responsibility of present generations to the future.[17] One of the most pressing questions is this: How will geoengineering define human beings as "makers" of the climate?[18] This is a profound philosophical, moral, and theological question, and religion is at the heart of how many individuals and most cultures explore such questions. For instance, some religious communities might justify geoengineering as a form of "stewardship," while others might consider it a prideful attempt to deny the limits of our humanity. Churches, synagogues, mosques, and temples often serve as the heart of the public square for the faithful on both sides of an issue, and thus religious communities take on the role of disseminators of information—or misinformation.

RELIGION CHALLENGES OR JUSTIFIES SCIENTIFIC AUTHORITY

Public discussions of science inevitably intersect with religious values and worldviews. Religious institutions may have stood in the way of science in the past, but this calls for a more careful engagement between the two in the present. Properly incorporated into the discourse, religion can help the wider public engage the science of climate change as it is currently understood, and geoengineering as humanity might conduct it.

One reason that discussions of geoengineering have paid little attention to religion to date is that the scientists and engineers involved in these discussions have understood religion to be outside their area of expertise. The complete separation of religion and science into distinct and nonoverlapping spheres is often believed to be the best way to keep peace between them. However, Templeton Prize winner and theologian Ian Barbour notes that separation is but one possible relationship between religion and science; it is also possible to understand the two as competitive, harmonious, or complementary.[19]

When it comes to climate change, most media attention has focused on competition between religion and science, as Joe Barton's quote and its inclusion in *Time* magazine's list of his "gaffes" demonstrate. As sociologist of religion Laurel Kearns observes, some who deny the problem of climate change do so out of a faith in capitalism rather than in God, trusting markets to respond to any issues that arise and distrusting the governments that sponsor climate science and would be necessary for a coordinated response.[20] When it begins to shape beliefs, behaviors, and societies, faith in markets can also be seen as religious.[21] Indeed, when such faith behaves like a religion but is not named as such, it can powerfully and subtly shape debate, leading people to dismiss the consensus view on climate change. These are uses of religion in opposition to science, and they undermine scientific literacy and authority.

On the other hand, many scientists have sought to engage religion cooperatively, believing that an informed and scientifically literate public is essential for reasoned action on an issue like climate change.[22] For example, astrophysicist and science advocate Carl Sagan led a group of scientists who made a plea for partnership with religion, writing: "Efforts to safeguard and cherish the environment need to be infused with a vision of the sacred. At the same time, a much wider and deeper understanding of science and technology is needed. If we do not understand the problem, it is unlikely we will be able to fix it. Thus, there is a vital role for religion and science."[23] This vision of science and religion partnering and complementing one another to address environmental threats is particularly relevant to geoengineering. Science is absolutely necessary to understand what is happening to the planet and to determine what human beings can do to change processes already underway. Religious institutions and traditions need to be able to learn from science. But scientists also have something to learn as they engage the issues of value, priority, and trust raised by geoengineering—issues that are beyond scientific authority.

Even if geoengineering proves to be feasible, it raises questions: Which human institutions have the authority to manage and regulate climate engineering? Who will decide how to value the nonhuman species that will be negatively impacted by recalibrating the climate for human life? What is the meaning of human existence, such that the technological humanization of the atmosphere is understood to be acceptable—or even laudable? To approach these questions, the sort of collaboration that Sagan and his coauthors called for is necessary: religious leaders and scientists working together with a recognition of their complementary expertise and authority.

RELIGION FRAMES GEOENGINEERING

Sacred narratives and theological metaphors already play a role in framing geoengineering, setting the terms and the norms of debate and discussion. As frames for understanding, religious narratives and symbols are provocative for believers and nonbelievers alike.

Alexa Spence and Nick Pidgeon, United Kingdom–based researchers in social and environmental psychology, offer an excellent description of a frame: "A frame allows complex issues to be pared down and for some aspects of that issue to be given greater emphasis than others in order that particular audiences can rapidly identify why an issue may be relevant to them."[24] In other words, a frame presents a model for how to understand, and therefore respond to, reality. Because religion engages questions of fundamental meaning and value, its metaphors and narratives are unusually adept at providing ideological frames.

One religious frame in contemporary debates provocatively asserts that climate engineering is an effort to "play God." Inherent in this metaphor is the religious conception of a tiered cosmos, wherein God or gods are associated with the heavens, while human beings belong on Earth. When people "play God" by tampering with the climate, they set foot on a larger playing field, taking on the role of the divine.

In this frame, whether geoengineering is appropriate or advisable depends upon whether one believes human beings must remain humble before the cosmos or should claim a more directive position as the dominant species on the planet. The frame presents possible arguments about geoengineering on a continuum: At one extreme, the idea of "playing God" condemns human hubris, while on the other, it is a theological justification of human skill.[25] Can—and should—human beings usurp authority from the gods and manage the atmosphere? Science writer Mark Lynas answers positively, referencing the Book of Genesis and a personal revelation when he suggests that geoengineering presents an opportunity to stave off environmental catastrophe: "Playing God is good for the planet."[26] In contrast, philosopher Clive Hamilton skeptically asks, "Is it wise to try to play God with the climate?"[27]

While "playing God" is the most obvious example of religion influencing the framing of geoengineering, there are others. The discourse of a coming "climate apocalypse" appeals to religious language to describe geoengineering, the problems it is attempting to solve, or both.[28] More indirectly, religious belief often orients the ideals that frame this debate: hope that human beings can live harmoniously in nature without technological intervention resonates with stories of Eden shared by the three major monotheistic religions, while hope that human beings can innovate and engineer a way out of environmental problems is consistent with teachings in the same religions that God gave human beings authoritative "dominion" over other creatures and the Earth as a whole.

Not every participant in discussions about geoengineering is directly influenced by religious framing, but any public discussion will encounter religious frames. Thus, participants in these debates must pay attention to religious language, religious images, and religious narratives that shape human capacity to interact with the world.

RELIGION OFFERS A VOCABULARY FOR MORAL DEBATE

Religions shape societies' and individuals' moral imaginations; so, as debates about geoengineering wrestle explicitly with moral challenges, religious vocabulary can offer an important set of resources. Of course, religions are not the only traditions and institutions that identify moral problems and offer language to discuss them. However, religions do possess vocabulary and means for such discussion that have been tested over millennia. For that reason, Christian theologian Mark Wallace proclaims that "only religion . . . has the moral authority and symbolic potency to break our shared carbon addiction cycle by motivating us to look beyond our private self-interest and to the greater good of the planet itself."[29]

In other words, tempered by time and familiarity, the ethical imagination conveyed through many religious traditions is relevant, evocative, and powerful. Religion often poses moral danger and virtue in ways that persuade not only intellectually but also emotionally. For example, Muslim activist Ibrahim Abdul-Matin calls on people of faith to turn away from fossil fuels by deeming them "energy from Hell" in contrast to the "energy from Heaven" provided by sun and wind.[30] This not only frames the debate about energy sources powerfully, it offers an incentive for a profound change, which scholars of religion would call conversion. Along similar lines, Christian theologian Cynthia Moe-Lobeda suggests that climate change must be understood as "sin," and that no response to it will be complete without "profound lament for the ways in which our lives unwittingly endanger Earth's life-system and vulnerable neighbors far and near."[31]

Such judgments of hellfire and sin are not dismissive and need not close off hope or conversation. Adbul-Matin's faith teaches that disobedience to Allah—service to Hell rather than Heaven—is a perennial temptation, but can be overcome by an appreciation for, and engagement with, God's creation. Moe-Lobeda emphasizes that Christian worship services incorporate tools for lament, which can help the community to recognize and redress its failings. Along similar lines, South African theologian Ernst Conradie argues that the global community can learn from the way Christians in his nation responded to the failings of apartheid—communally facing, lamenting, and redressing past wrongs. Climate change, he argues, calls for communal penance of the type demonstrated by the Truth and Reconciliation Commission.[32]

Such discussions of penance and conversion are thus far missing from moral conversation about geoengineering. Consider, for example, the crucial debate about whether geoengineering creates a moral hazard, such that pursuing adaptations will weaken efforts to mitigate climate change. Alan Robock makes this case, suggesting that the promise of an "easy technological fix for global warming" will inevitably reduce national and international support for efforts to reduce consumption and reform energy infrastructure.[33] Keith and his colleagues disagree, insisting that the most profound danger is inaction and that climate engineering "may be the only human response that can fend off rapid and high-consequence climate impacts."[34] At its roots, this is a disagreement about the nature of the problem geoengineering attempts to solve. Robock seems to assume that the core problem is the hubris that led to climate change, and so is cautious about humanity taking on more power and authority. Keith and his colleagues seem to assume that the core problem is inaction, and so resist any effort to hold back new and innovative responses to the problem. Interestingly, these assumptions are not stated; the moral problem that geoengineering seeks to solve is assumed rather than stated.

Seen through the lens of Christian theological ethics, these perspectives are calling for repentance from different sins: Robock is concerned about the sin of prideful action, while Keith and his colleagues are concerned about the sin of slothful inaction. Of course, a dialogue with religion will not resolve this dispute, but it does offer terminology and perspectives to make the stakes more explicit. Is the human species facing the greatest danger from extending ourselves too far to manage the world, or from neglecting our responsibilities to one another and the rest of the planet? In a world of anthropogenic atmospheric change, which sin requires more urgent conversion: pride or sloth? Discussing that question will advance the debate about whether geoengineering is a moral response to climate change or an immoral continuation of the same mistakes that caused climate change in the first place.

ENGAGING RELIGION

In his final book, *Where Do We Go from Here: Chaos or Community?*, civil rights activist and Baptist preacher Martin Luther King Jr. called on his readers to "work passionately and indefatigably to bridge the gulf between our scientific progress and our moral progress." He wondered at the accomplishments of science but noted a "moral and spiritual lag" behind such developments, cautioning that "when scientific power outruns moral power, we end up with guided missiles and misguided men."[35]

Like nuclear technology, geoengineering presents a range of astonishing scientific possibilities that deserve careful study and careful discussion. As with all scientific proposals, they also require careful moral and spiritual reflection. How do humans interpret our Promethean aspirations, our dreams of tending the garden or of using the tools of the gods? How might our actions lay claim on our souls? Such questions inevitably emerge in the midst of human world-transformation.

When discussing whether and how it is acceptable to "play God" with the climate, it is essential to recognize that this is, in part, a religious game. For many—citizens and policy makers, scientists and engineers—religion is not simply a matter of abstract interest, but a direct influence on how to make sense of the world and act within it. When playing God, it is necessary to determine how that matters, why it matters, and to whom it matters. If research into geoengineering is necessary (something on which most commentators seem to agree), then religion must have a place at the table. This does not mean ceding the discussion entirely to religion—the strategy suggested by Rep. Joe Barton. Rather, it means finding a way to make religion a part of the multifaceted, interdisciplinary conversation about climate change.

Of course, religion is a wildly diverse category encompassing a broad array of human experiences, systems, and organizations. It would not be practical for scientists and engineers to consult with all religious people, or even all religious leaders, in the course of their work. However, since geoengineering is currently discussed most prominently by academic experts, it is important that experts who study and reflect on religious belief—scholars of theology, ethics, and religious studies—be part of the discussion. Such scholars offer many of the unique perspectives and tools of religion, but also develop comparative and critical perspectives on religious traditions. That perspective will advance not only experts' discussions of geoengineering but also the vital work of bringing those discussions to a wider public.

NOTES

1. This chapter was originally published in *Bulletin of the Atomic Scientist* 70:3 (2014). We are grateful to the *Bulletin* for the right to republish it here, and also grateful to Dawn Stover, the editor there who worked to develop and refine this piece.

2. Jay Newton-Small, and Katy Steinmetz, "Eight More Deep Thoughts from Rep. Joe Barton," *Time*, June 18 (2010), http://content.time.com/time/politics/article/0,8599,1997963,00.html.

3. The British Royal Society, *Geoengineering the Climate: Science, Governance, and Uncertainty* (London: Royal Society Reports, 2009), 1.

4. In response to the increased interest in geoengineering, several useful overviews have been published, including Clive Hamilton, *Earthmasters: The Dawn of the Age of Climate*

Engineering (New Haven, CT: Yale University Press, 2013); David W. Keith, "Geoengineering the Climate: History and Prospect," *Annual Review of Energy and the Environment* 25 (2000); M. Granger Morgan and Katharine Ricke, *Cooling the Earth through Solar Radiation Management* (Geneva: Interantional Risk Governance Council, 2010); and Naomi Vaughan and Timothy Lenton, "A Review of Climate Geoengineering Proposals," *Climatic Change* 109: 3–4 (2011). For an overview of the ethical debate surrounding geoengineering, see Christopher J. Preston, "Ethics and Geoengineering: Reviewing the Moral Issues Raised by Solar Radiation Management and Carbon Dioxide Removal," *WIREs Climate Change* 4:1 (2013).

5. David W. Keith, Edward Parson, and M. Granger Morgan, "Research on Global Sun Block Needed Now," *Nature* 463 (2010).

6. Dan Whaley, and Margaret S. Leinen, "Designing a Geoengineering Research Agenda Should Be a Group Effort." July 28. http://thebulletin.org/ has-time-come-geoengineering/designing-geoengineering-research-agenda-should-be-group-effort.

7. David G. Victor, "On the Regulation of Geoengineering," *Oxford Review of Economic Policy* 24, no. 2 (2008).

8. Kennette Benedict, "Existential Threats, Fast and Slow," July 18 (2013), http://thebulletin.org/existential-threats-fast-and-slow.

9. See Wiley Carr, Ashley Mercer, and Clare Palmer, "Public Concerns about the Ethics of Solar Radiation Management," in *Engineering the Climate: The Ethics of Solar Radiation Management* (Lanham, MD: Lexington, 2012); Wiley Carr et al., "Public Engagement on Solar Radiation Management and Why It Needs to Happen Now," *Climatic Change* 121 (2013).

10. Alan Robock, "20 Reasons Why Geoengineering May Be a Bad Idea," *Bulletin of the Atomic Scientists* 64, no. 2 (2008).

11. Frank Newport, *God Is Alive and Well: The Future of Religion in America* (New York: Gallup, 2012).

12. Pew Forum on Religion & Public Life, "The Global Religious Landscape," (2012), http://www.pewforum.org/files/2014/01/global-religion-full.pdf.

13. See for example Dieter Gerten and S. Bergmann, eds. *Religion in Environmental and Climate Change: Suffering, Values, Lifestyles* (New York: Continuum, 2012); Michael S. Northcott, *A Moral Climate: The Ethics of Global Warming* (Maryknoll, NY: Orbis Books, 2007); Jame Schaefer, *Confronting the Climate Crisis: Catholic Theological Perspectives* (Milwaukee, WI: Marquette University Press, 2011); John Stanley, David Loy, and Dorje Gyurme, *A Buddhist Response to the Climate Emergency* (Boston: Wisdom Publications, 2009).

14. See for example Anders Biel and Andreas Nilsson, "Religious Values and Environmental Concern: Harmony and Detachment," *Social Science Quarterly* 86, no. 1 (2005); Andrew Greeley, "Religion and Attitudes toward the Environment," *Journal for the Scientific Study of Religion* 32, no. 1 (1993); James L. Guth et al., "Faith and the Environment: Religious Beliefs and Attitudes on Environmental Policy," *American Journal of Political Science* 39, no. 2 (1995); Darren E. Sherkat and Christopher G. Ellison, "Structuring the Religion-Environment Connection: Identifying Religious Influences on Environmental Concern and Activism," *Journal for the Scientific Study of Religion* 46, no. 1 (2007).

15. David C. Barker and David H. Bearce, "End-Times Theology, the Shadow of the Future, and Public Resistance to Addressing Global Climate Change," *Political Research Quarterly* 66, no. 2 (2013).

16. Forrest Clingerman, "Between Babel and Pelagius: Religion, Theology, and Geoengineering," in *Engineering the Climate: The Ethics of Solar Radiation Management*, ed. Christopher J. Preston (New York: Lexington Books, 2012).

17. See Forrest Clingerman, "Geoengineering, Theology, and the Meaning of Being Human," *Zygon* 49, no. 1 (2014); and Adam Corner, Nick Pidgeon, and Karen Parkhill, "Perceptions of Geoengineering: Public Attitudes, Stakeholder Perspectives, and the Challenge of 'Upstream' Engagement," *WIREs Climate Change* 3, no. 5 (2012).

18. Maialen Galarraga and Bronislaw Szerszynski, "Making Climates: Solar Radiation Management and the Ethics of Fabrication," in *Engineering the Climate: The Ethics of Solar Radiation Management* (Lanham, MD: Lexington, 2011).

19. Ian G. Barbour, *Religion and Science: Historical and Contemporary Issues* (San Francisco: HarperSanFrancisco, 1997).

20. Laurel Kearns, "Cooking the Truth: Faith, Science, the Market, and Global Warming," in *Ecospirit: Religions and Philosophies for the Earth*, ed. Laurel Kearns and Catherine Keller (New York: Fordham University Press, 2007).

21. David Loy, "The Religion of the Market," *Journal of the American Academy of Religion* 65, no. 2 (1997).

22. Paul R. Ehrlich and Anne H. Ehrlich, *Betrayal of Science and Reason: How Anti-Environmental Rhetoric Threatens Our Future* (Washington, DC: Island Press, 1996); Albert Gore, *The Assault on Reason* (New York: Penguin Press, 2007).

23. Carl Sagan, "An Open Letter to the Religious Community," (1990).

24. Alexa Spence and Nick Pidgeon, "Framing and Communicating Climate Change: The Effects of Distance and Outcome Framing Manipulations," *Global Environmental Change* 20, no. 4 (2010), 657.

25. See Clingerman, "Between Babel and Pelagius: Religion, Theology, and Geoengineering."; C. A. J. Coady, "Playing God," in *Human Enhancement*, ed. Julian Savulescu and Nick Bostrom (New York: Oxford University Press, 2009); and Willem B. Drees, "Playing God? Yes! Religion in the Light of Technology," *Zygon* 37, no. 3 (2002).

26. Mark Lynas, "Geoengineering Nuclear Power, and Climate Change: Playing God Is Good for the Planet," *The Telegraph*, July 12 (2011).

27. Clive Hamilton, "Geoengineering: Our Last Hope, or a False Promise?" *New York Times*, May 27 (2013).

28. Forrest Clingerman and Verna Ehret, "Hope and Fear: The Theological Side of Framing Environmental Change," *Ethics, Policy, and Environment* 16, no. 2 (2013); Cheryl Hall, "What Will It Mean to Be Green? Envisioning Positive Possibilities without Dismissing Loss," *Ethics, Policy, and Environment* 16, no. 2 (2013); Mike Hulme, "The Conquering of Climate: Discourses of Fear and Their Dissolution," *Geographical Journal* 174, no. 1 (2008); Stefan Skrimshire, *Future Ethics: Climate Change and Apocalyptic Imagination* (New York: Continuum, 2009).

29. Mark I. Wallace, *Green Christianity: Five Ways to a Sustainable Future* (Minneapolis: Fortress Press, 2010), 1, 2–3.

30. Ibrahim Abdul-Matin, *Green Deen: What Islam Teaches about Protecting the Planet* (San Francisco, CA: Berrett-Koehler Publishers, 2010).

31. Cynthia D. Moe-Lobeda, *Resisting Structural Evil: Love as Ecological-Economic Vocation* (Minneapolis, MN: Fortress Press, 2013), 26, 1–2.

32. Ernst M. Conradie, *An Ecological Christian Anthropology: At Home on Earth?* (Burlington, VT: Ashgate, 2005).

33. Robock, "20 Reasons Why Geoengineering May Be a Bad Idea," 17.

34. Keith, Parson, and Morgan, "Research on Global Sun Block Needed Now," 426.

35. Martin Luther King, *Where Do We Go from Here: Chaos or Community?* (London: Hodder & Stoughton, 1967), 17, 2–3.

BIBLIOGRAPHY

Abdul-Matin, Ibrahim. *Green Deen: What Islam Teaches About Protecting the Planet*. San Francisco, CA: Berrett-Koehler Publishers, 2010.

Barbour, Ian G. *Religion and Science: Historical and Contemporary Issues*. San Francisco: HarperSanFrancisco, 1997.

Barker, David C., and David H. Bearce. "End-Times Theology, the Shadow of the Future, and Public Resistance to Addressing Global Climate Change." *Political Research Quarterly* 66, no. 2 (2013): 267–79.

Benedict, Kennette. "Existential Threats, Fast and Slow." July 18 (2013). http://thebulletin.org/existential-threats-fast-and-slow.

Biel, Anders, and Andreas Nilsson. "Religious Values and Environmental Concern: Harmony and Detachment." *Social Science Quarterly* 86, no. 1 (2005): 178–91.

The British Royal Society. *Geoengineering the Climate: Science, Governance, and Uncertainty*. London: Royal Society Reports, 2009.

Carr, Wiley, Ashley Mercer, and Clare Palmer. "Public Concerns about the Ethics of Solar Radiation Management." In *Engineering the Climate: The Ethics of Solar Radiation Management*, 169–86. Lanham, MD: Lexington, 2012.

Carr, Wiley, Christopher J. Preston, Laurie Yung, Bronislaw Szerszynski, David W. Keith, and Ashley M. Mercer. "Public Engagement on Solar Radiation Management and Why It Needs to Happen Now." *Climatic Change* 121 (2013): 567–77.

Clingerman, Forrest. "Between Babel and Pelagius: Religion, Theology, and Geoengineering." In *Engineering the Climate: The Ethics of Solar Radiation Management*, edited by Christopher J. Preston, 201–19. New York: Lexington Books, 2012.

———. "Geoengineering, Theology, and the Meaning of Being Human." *Zygon* 49, no. 1 (2014): 6–21.

Clingerman, Forrest, and Verna Ehret. "Hope and Fear: The Theological Side of Framing Environmental Change." *Ethics, Policy, and Environment* 16, no. 2 (2013): 152–55.

Coady, C. A. J. "Playing God," In *Human Enhancement*, edited by Julian Savulescu and Nick Bostrom, 15, 5–180. New York: Oxford University Press, 2009.

Conradie, Ernst M. *An Ecological Christian Anthropology: At Home on Earth?* Burlington, VT: Ashgate, 2005.

Corner, Adam, Nick Pidgeon, and Karen Parkhill. "Perceptions of Geoengineering: Public Attitudes, Stakeholder Perspectives, and the Challenge of 'Upstream' Engagement." *WIREs Climate Change* 3, no. 5 (2012): 451–66.

Drees, Willem B. "Playing God? Yes! Religion in the Light of Technology." *Zygon* 37, no. 3 (2002): 643–54.

Ehrlich, Paul R., and Anne H. Ehrlich. *Betrayal of Science and Reason: How Anti-Environmental Rhetoric Threatens Our Future*. Washington, DC: Island Press, 1996.

Galarraga, Maialen, and Bronislaw Szerszynski. "Making Climates: Solar Radiation Management and the Ethics of Fabrication." In *Engineering the Climate: The Ethics of Solar Radiation Management*, 221–35. Lanham, MD: Lexington, 2011.

Gerten, Dieter, and Sigurd Bergmann, eds. *Religion in Environmental and Climate Change: Suffering, Values, Lifestyles.* New York: Continuum, 2012.

Gore, Albert. *The Assault on Reason*. New York: Penguin Press, 2007.

Greeley, Andrew. "Religion and Attitudes toward the Environment." *Journal for the Scientific Study of Religion* 32, no. 1 (1993): 19–28.

Guth, James L., John C. Green, Lyman A. Kellstedt, and Corwin E. Smidt. "Faith and the Environment: Religious Beliefs and Attitudes on Environmental Policy." *American Journal of Political Science* 39, no. 2 (1995): 364–82.

Hall, Cheryl "What Will It Mean to Be Green? Envisioning Positive Possibilities without Dismissing Loss." *Ethics, Policy, and Environment* 16, no. 2 (2013): 125–41.

Hamilton, Clive. *Earthmasters: The Dawn of the Age of Climate Engineering.* New Haven, CT: Yale University Press, 2013.

———. "Geoengineering: Our Last Hope, or a False Promise?" *New York Times*, May 27 (2013): A17.

Hulme, Mike. "The Conquering of Climate: Discourses of Fear and Their Dissolution." *Geographical Journal* 174, no. 1 (2008): 5–16.

Kearns, Laurel. "Cooking the Truth: Faith, Science, the Market, and Global Warming." In *Ecospirit: Religions and Philosophies for the Earth*, edited by Laurel Kearns, and Catherine Keller, 97–124. New York: Fordham University Press, 2007.

Keith, David W. "Geoengineering the Climate: History and Prospect." *Annual Review of Energy and the Environment* 25 (2000): 245–84.

Keith, David W., Edward Parson, and M. Granger Morgan. "Research on Global Sun Block Needed Now." *Nature* 463 (2010): 245–85.

King, Martin Luther. *Where Do We Go from Here: Chaos or Community?* London: Hodder & Stoughton, 1967.

Loy, David. "The Religion of the Market." *Journal of the American Academy of Religion* 65, no. 2 (1997): 275–90.

Lynas, Mark. "Geoengineering Nuclear Power, and Climate Change: Playing God Is Good for the Planet." *The Telegraph*, July 12 (2011).

Moe-Lobeda, Cynthia D. *Resisting Structural Evil: Love as Ecological-Economic Vocation*. Minneapolis, MN: Fortress Press, 2013.

Morgan, M. Granger, and Katharine Ricke. *Cooling the Earth through Solar Radiation Management: The Need for Research and an Approach to Its Governance*. Geneva: International Risk Governance Council, 2010.

Newport, Frank. *God Is Alive and Well: The Future of Religion in America*. New York: Gallup Press, 2012.

Newton-Small, Jay, and Katy Steinmetz. "Eight More Deep Thoughts from Rep. Joe Barton." *Time*, June 18 (2010): http://content.time.com/time/ politics/article/0,8599,1997963,00.html. Accessed Jan 17, 2016.

Northcott, Michael S. *A Moral Climate: The Ethics of Global Warming*. Maryknoll, NY: Orbis Books, 2007.

Pew Forum on Religion & Public Life. "The Global Religious Landscape." (2012). http://www.pewforum.org/files/2014/01/global-religion-full.pdf.

Preston, Christopher J. "Ethics and Geoengineering: Reviewing the Moral Issues Raised by Solar Radiation Management and Carbon Dioxide Removal." *WIREs Climate Change* 4, no. 1 (2013): 23–37.

Robock, Alan. "20 Reasons Why Geoengineering May Be a Bad Idea." *Bulletin of the Atomic Scientists* 64, no. 2 (2008): 14–18.

Sagan, Carl. "An Open Letter to the Religious Community." (1990).

Schaefer, Jame. *Confronting the Climate Crisis: Catholic Theological Perspectives*. Milwaukee, Wis.: Marquette University Press, 2011.

Sherkat, Darren E., and Christopher G. Ellison. "Structuring the Religion-Environment Connection: Identifying Religious Influences on Environmental Concern and Activism." *Journal for the Scientific Study of Religion* 46, no. 1 (2007): 71–85.

Skrimshire, Stefan. *Future Ethics: Climate Change and Apocalyptic Imagination*. New York: Continuum, 2009.

Spence, Alexa, and Nick Pidgeon. "Framing and Communicating Climate Change: The Effects of Distance and Outcome Framing Manipulations." *Global Environmental Change* 20, no. 4 (2010): 656–67.

Stanley, John, David Loy, and Dorje Gyurme. *A Buddhist Response to the Climate Emergency*. Boston: Wisdom Publications, 2009.

Vaughan, Naomi E., and Timothy M. Lenton. "A Review of Climate Geoengineering Proposals." *Climatic Change* 109, no. 3–4 (2011): 745–90.

Victor, David G. "On the Regulation of Geoengineering." *Oxford Review of Economic Policy* 24, no. 2 (2008): 322–36.

Wallace, Mark I. *Green Christianity: Five Ways to a Sustainable Future*. Minneapolis: Fortress Press, 2010.

Whaley, Dan, and Margaret S. Leinen. "Designing a Geoengineering Research Agenda Should Be a Group Effort." July 28. Accessed October 3, 2015. http://thebulletin.org/has-time-come-geoengineering/designing-geoengineering-research-agenda-should-be-group-effort.

Chapter Two

From the Garden of Eden to Eden's Gardener?—Experiences from Dialogues with Religious Groups on Climate Engineering and Possible Implications for Transdisciplinarity

Thomas Bruhn, Stefan Schäfer, and Mark G. Lawrence[1]

THE RELEVANCE OF ENGAGING RELIGIOUS GROUPS ON THE TOPIC OF CLIMATE ENGINEERING

Climate change remains an urgent challenge for humanity. Beyond the primary approaches to addressing climate change, namely by reducing the emissions of anthropogenic greenhouse gases (mitigation) and by adapting to the environmental changes resulting from global warming (adaptation), proposals for intentionally intervening in the Earth system to counteract rising global mean temperatures are increasingly being discussed. These proposals, often summarized under the term "climate engineering" (also "geoengineering" or "climate interventions"), originally emerged from Earth system science, especially from atmospheric science.[2] As discussed in the introduction to this volume, the term climate engineering encompasses a broad spectrum of proposals, which can mostly be grouped under two categories: either removing greenhouse gases like carbon dioxide from the atmosphere (generally "carbon dioxide removal" or more broadly "greenhouse gas removal"), or modifying the reflectivity of the Earth to increase the amount of sunlight that is reflected back to space (generally called "solar radiation management" or "albedo modification").[3]

Over the last decade, the idea of climate engineering has increasingly provoked controversial discussions about scientific and technical issues, as well as about the complexities of relevant societal issues such as governance, international law, and ethics. In this context, it is not unlikely that religious communities may come to develop a strong interest in the topic of climate engineering. For example, religious communities may perceive discussions of climate engineering as raising challenging questions about the relationship between human beings and the nonhuman world, or about the protection of poor communities and ecosystems from the impacts of climate change.[4] Some have alluded to an understanding of climate engineering as "playing God."[5] How individuals and groups come to form specific understandings of climate engineering, based in part on how religious beliefs and institutions contribute to shaping discussions and potentially decision making, thus represents relevant avenues of scholarly inquiry.

An example of this framing is the concept of "good stewardship" of the Earth, which is applied frequently in the Judeo-Christian context of discussions of climate change in North America. As Steffen et al. have pointed out, the present generation is "the first generation with widespread knowledge of how our activities influence the Earth System, and thus the first generation with the power and the responsibility to change our relationship with the planet."[6] So, the question arises as to how to interpret responsible stewardship in the Anthropocene, particularly as the severity of climate change may lead societies to consider more drastic options like climate engineering.

In addition to scholarly research, and taking into account that nonacademic perspectives—such as those of religious groups—can significantly contribute to shaping the course of scientific and technological developments, early engagement with nonacademic perspectives may contribute to creating a shared understanding of what is at stake in emerging areas of scientific investigation and technology development. In order to engage with these nonacademic perspectives, an interdisciplinary climate engineering research group (called the Climate Engineering Research Microcosm) at the IASS Potsdam has sought an open exchange with civil society and religious groups as part of the ambition to develop a transdisciplinary approach to climate engineering research. An important motivation behind this was to ensure that the critical issues around climate engineering are discussed with a variety of perspectives before decisions are made. This involved the desire to include stakeholders that were so far unaware of the topic, but that, if not involved early on, might come to develop a strong interest in the debate at later stages. Without early engagement, there is the risk that stakeholders may feel presented with a *fait accompli*, which may provoke principled opposition and significantly hinder open exchange and dialogue.

In this chapter, we describe our experiences so far and reflect on the future perspectives of this exchange in the context of transdisciplinarity. In particular, we will summarize the insights obtained from the exchange with Christian groups and give an overview of the issues around climate engineering that were identified to be of special relevance for them.

TRANSDISCIPLINARITY—A CHALLENGE FOR ACADEMIC PRACTICE

The concept of transdisciplinarity is still rather new and far from well-established in the academic community, but it has been increasingly discussed and applied since the early 2000s.[7] Particularly in the context of sustainability, the complex and interdependent nature of societal problems has led some to conclude that these problems can often not be addressed adequately by segmented thinking and problem solving within disciplinary or sectoral boundaries.[8] Transdisciplinarity recognizes that the complexity of societal problems requires integrating academic with nonacademic forms of knowledge in a collaborative process with societal partners (stakeholders). Various definitions of transdisciplinarity are in use. Based on an extensive literature review, Jahn et al. have proposed a working definition of transdisciplinarity as "a reflexive research approach that addresses societal problems by means of interdisciplinary collaboration as well as the collaboration between researchers and extra-scientific actors; its aim is to enable mutual learning processes between science and society; integration is the main cognitive challenge of the research process."[9]

In addition to the lack of clarity in the definition of transdisciplinarity, the theory and practice of transdisciplinarity are also not always congruent, owing also in part to the early stage of methodological development in this area.[10]

It is important to acknowledge that transdisciplinarity is simultaneously an attitude and a mode of action, meaning that beyond its methodological conception it also encompasses the practice of a certain mind-set as described later. As it recognizes the relevance of nonscientific forms of knowledge and actions in addressing societal challenges, the vision of transdisciplinarity implies that academic and nonacademic perspectives meet and exchange on an eye-to-eye level. This is particularly important in the initial phase of exchange between academics and nonacademic stakeholders. In this phase it is crucial that the outcome of the process is not predefined.[11] Rather, the first phase of transdisciplinary processes needs to aim for developing a truly joint understanding of a "real world" problem before a concrete desired outcome can then be defined together. Stakeholders and academics may have to acknowledge that a problem may exhibit different facets from the respective perspectives of the different involved partners. In order to set up a joint process of developing strategies and solutions together, it is therefore crucial to firstly come to a genuinely shared understanding of the respective problem, a phase that Jahn et al. have called "problem transformation," referring to the fact that the various different initial understandings of a problem need to be transformed into one joint problem understanding that is shared by all involved partners.[12] Coming to this shared understanding is essential as it lays the basis for further generating new knowledge and solutions together in the later stages of the transdisciplinary process.

Throughout their entire duration, transdisciplinary processes therefore raise several challenges regarding communication and exchange formats between academics and nonacademic partners. In the first phase of a joint "problem transformation," a principal task is to facilitate genuine dialogue and to elaborate new shared language, logic, and concepts.[13] The transdisciplinary involvement of extrascientific actors goes beyond one-directional information or consultation on a predefined agenda. It aims at establishing a form of collaboration that empowers participants to actually influence the agenda and to question and possibly modify the dominant structures and understandings that guide processes of cogenerating knowledge and meaning.[14] This can require mediation to define collectively what could and should be done and to ensure that integrative forms of communication receive particular attention.[15] Specifically, transdisciplinarity calls for "humility, openness to others, a contextualization of our own knowledge and a willingness to engage with and be moved by others."[16] Jahn et al. emphasize that transdisciplinarity "requires an uncommon willingness of individual scientists to learn and to think outside of the disciplinary box. This willingness, in return, crucially depends on the extent to which individual interests are recognized and supported."[17]

These ambitions signal that many see transdisciplinarity as being more than an academic concept or methodology. Rather, they see transdisciplinarity as challenging the involved individuals to practice a certain "transdisciplinary mind-set" that encompasses the characteristics mentioned above, with the goal of allowing genuine engagement between groups whose perspectives may be based on very different rationalities, assumptions, and values.

In the following, we will describe our experiences with a dialogue series on climate engineering and its relationship to the ambitions and principles of transdisciplinarity. First we will describe the format ("climate engineering dialogue series") that we established for the exchange with nonacademic stakeholders and provide some examples of how this format worked in practice. More specifically, we will then summarize the experience we made in the exchange with Christian perspectives and the concrete impulses regarding climate engineering that resulted from this exchange. We will conclude with some reflections on the lessons learned with respect to transdisciplinarity and strategies for communicating about the topic of climate engineering with religious stakeholders.

THE CLIMATE ENGINEERING DIALOGUE SERIES OF THE IASS POTSDAM

Building on the idea of transdisciplinarity, the Institute for Advanced Sustainability Studies (IASS) in Potsdam, Germany, was established in 2009 with the ambition to not only perform research on sustainability science but to also act as a platform and convener for strategic dialogues in order to support transformative processes in society toward sustainability.[18]

At the IASS, the interdisciplinary "Climate Engineering Research Microcosm" approaches the topic of climate engineering in a holistic way. Its current and past members include academics from a very broad range of fields like physics, atmospheric science, political science, law, philosophy, psychology, economics, and geography. It is important to emphasize that the IASS does not hold an official position on climate engineering. Rather, coupled to its research and policy activities, the IASS provides a platform for facilitating a learning process about climate engineering that encompasses as many relevant perspectives as possible, in order to allow for an inclusive assessment and for research and policy making to proceed in close exchange with societal stakeholders.

Since the spring of 2013, the IASS has organized a series of dialogue events on climate engineering with German civil society organizations (CSOs) and representatives of religious groups. At that point in time, climate engineering was not prominently discussed in the German public. Few public media outlets had reported about the topic, and most stakeholder groups involved in the dialogue series stated that they had first heard of the topic when they were approached by the IASS. Often, a long time (usually several months) passed between the initial contact and the actual dialogue meeting because the topic was very low on the agenda of the dialogue partners.

Idea and Goals

The initial idea for the dialogue series was to better connect the work of the Climate Engineering Research Microcosm to relevant stakeholders in society. On the one hand, it was intended to inform interested stakeholders about the IASS's work on climate engineering. At the same time, however, it was also intended to provide an opportunity for identifying which aspects stakeholders considered to be the most relevant, so that the members of the Microcosm could take this into consideration in their work and in their choice of research topics. Against this background, it became a primary goal of the dialogues to listen rather than to lecture, and to develop mutual trust as a basis for future exchange on potentially more specific issues.[19] In this sense, it also reflects the notion of a deliberative public engagement, as, for example, called for by Carr et al.[20] It is important to emphasize that the dialogue series was not designed for the purpose of performing social scientific research on stakeholder perceptions or attitudes toward climate engineering.[21]

Format and Setting

In order to pursue the intentions described above, in the dialogue events held thus far we have found that it was essential to provide enough time for open discussion and a comfortable, informal setting. Meetings were usually scheduled for three hours and were designed to provide space for open and inclusive discussion, rather than formal presentations. Generally, the participants sat around one table that was as small as possible, without a gap in the middle, and without laptops on the table. There was no prearranged order of seating. However, it was taken care that participants from IASS and the stakeholder group generally did not sit as two parties on separate sides of the table but rather mixed among the two groups. It was also ensured that the number of IASS participants generally did not exceed the number of participants from the respective stakeholder group.

We did not include an external mediator, as proposed, for example, by Klein.[22] The IASS acted as a "position-neutral" platform in the climate engineering debate and only one stakeholder group was invited to a meeting at a time, so we did not anticipate highly contentious discussion that might require being mediated.

The meetings were held under Chatham House Rules. Therefore all expressed opinions reflected personal opinions but not official positions of the institutions taking part in the dialogue. A typical agenda is displayed in table 2.1.

Experiences and Anecdotes

Often, participants entered the meetings skeptically. Most had not been familiar with the topic before they were approached by the IASS. Several participants reported that they prepared for the meeting by gathering some quick information from the German news media and had gained the impression that scientists saw climate engineering as a potentially "quick and easy way out of the climate problem." With this impression in mind, many participants had expected to be lectured similarly from the IASS and were prepared to argue against this view. One group had even prepared a quick position paper in order to be prepared for a confrontational discussion.

The round of personal introductions (see table 2.1) was used by the IASS participants to emphasize that the purpose of the meeting was to openly discuss issues in climate engineering and to listen to the participants' potential concerns. Several guests stated afterward that their expectations of confrontation and of "being sold a new scientific solution" had been "disap-

Table 2.1.

Time	**Item**	**Explanations/comments**
15:00	Get together	Informal welcome and serving coffee
15:05	Introduction and welcome	Background and motivation for this meeting
15:10	Round of personal introductions	Personal background and experience with climate engineering Hopes and expectations for this meeting?
15:30	Impulse CSO (optional)	Opportunity for the CSO to present themselves in free format
15:45	Presentation IASS	direct comprehensive questions on the presentation but no in-depth discussion about specific aspects
16:15	Round of personal statements on the topic	No discussion; opportunity for everyone to express personal impressions and opinions after presentation
16:45	Discussion	Common basic tenor on climate engineering? Impulses and recommendations for research? Recommendations for concrete activities?
17:45	Feedback round	Personal comments and feedback on meeting format
18:00	Farewell	

pointed" in a positive sense. So, this round of introductions turned out to be extremely valuable for achieving a constructive atmosphere and good dynamics of the meeting.

Often this round took longer than scheduled. However, it was clearly decided not to interrupt this personal check-in in order to lay the basis for a trustful discussion in which it is clear that all individual perspectives are valued by everyone in the room.

The impulse from the CSO gave the guests the opportunity to speak first and to express their perspective generally as an organization and, if desired, more specifically on the topic of climate engineering.

The presentation by the IASS provided a brief introduction and overview on the science and critical questions around climate engineering. Usually it was structured as follows:

- brief overview of climate change and current efforts toward mitigation and adaptation
- carbon dioxide removal

-framed as an approach to address the physical cause of climate change (GHG concentrations)

-overview of the range of technological suggestions

-details about a few selected techniques, for example, ocean iron fertilization, bioenergy with carbon capture and storage, and free air capture

- solar radiation management (a.k.a. albedo modification)

-framed as a suggestion to address the main symptom of climate change (the increase in global surface temperatures)

-overview of the technological suggestions

-more detailed description of selected techniques, especially stratospheric aerosol injections

- climate engineering

-framed as umbrella term for carbon dioxide removal and solar radiation management

-similarities and differences to related ideas like weather modification and terra-forming

-critical questions from various kinds of disciplines and perspectives, especially in terms of the social and ethical perspectives and the challenges of developing effective governance

-the interdisciplinary approach of Climate Engineering Research Microcosm focusing on an assessment of potentials, risks and side effects together with societal stakeholders

In the *round of personal statements* after the presentation, everyone (including participants from the IASS) was given the opportunity to express their personal impressions and feelings about the topic as a kickoff for the discussion. The statements were not discussed, but they cleared the air in the sense that everyone could express what urgently moved him or her. By doing this, a certain transparency was created with regard to the personal perspectives present in the room. Importantly, after this round of statements everyone in the room had spoken before the in-depth discussion started. Our impression was that this helped in creating an atmosphere in which effectively everyone participated actively and contributed to the discussion.

For the *discussion*, as explained above, the IASS participants emphasized that the main idea was to learn from the participants about their perspectives and to receive input from them. Because the discussion was opened this way, the concrete topics and focal points were largely defined by the participants.

Guiding questions brought into the discussion by the IASS moderator were:

- Which aspects do you consider to be most critical about climate engineering?
- Are there commonalities between the participants (guests and from the IASS) in their viewpoints on climate engineering?
- Which concrete impulses and recommendations do the participants want to provide to the IASS?
- Are there ideas or recommendations for specific activities?

Impacts and Follow-Up Activities

The main outcomes of the dialogues were the development of a trusting relationship to the participants and an increased understanding and sensitivity within the Climate Engineering Research Microcosm for stakeholder perspectives and opinions on the topic of climate engineering.

Several developments emerged as follow-ups to the meetings, especially interaction on preparation of position papers by the participating groups, which took on various forms, including:

- asking the IASS for direct critical comments on the text, focusing on the accuracy of the scientific arguments underlying the position statements;
- inviting researchers from the Climate Engineering Research Microcosm to take part in an in-depth briefing of a working group preparing a position paper on climate engineering;
- setting up an internal working group on climate engineering and developing a statement paper together with the IASS dialogue organizer, which was presented in a joint press conference.

DETAILED EXPERIENCES FROM THE EXCHANGE WITH CHRISTIAN PERSPECTIVES

In the exchange with religious groups, we often experienced skepticism about climate engineering. Similar to our experience with nonreligious stakeholders, this skepticism focused more on solar radiation management, while carbon dioxide removal was seen with greater interest and support, as also described by several earlier publications on stakeholder engagement on climate engineering.[23] Also, there was serious concern that a broad discussion about climate engineering could serve industrial interests that already prevent effective mitigation efforts.

In the following, we would like to describe several aspects of the discussion that were specific to the individuals who took part in the dialogue. The participants referred to in this section are all from a Roman Catholic background. This was not due to a preference of a specific religious perspective over others, but happened to be the religion for which the most intense and highest-level contacts were established. Similar high-level and intense dialogues with representatives from other religions are being sought out. As described earlier, we would like to note that the statements made in this dialogue process reflect personal opinions of the participants, but not official positions of any Christian institution. The reflections presented in this section represent an experience report rather than the results of an in-depth scientific analysis. This is particularly because the meetings were primarily designed to facilitate genuine open dialogue and establish a trusting relationship, without any constraints imposed by a formal framework that is necessary for enabling robust social scientific inquiry.

We identified three main focal areas of this discussion:

1. The role of humans on Earth and human responsibility for nature,
2. Interpretation and implications of this general understanding with respect to assessing climate engineering, and
3. Concrete comments or recommendations with respect to climate engineering research.

THE ROLE OF HUMANS ON EARTH AND HUMAN RESPONSIBILITY FOR NATURE

An important basis for addressing the concrete issue of climate engineering seemed to be the fundamental interpretation of how Christianity represents the role of humans on Earth.

Participants emphasized that stories from the Bible can provide central insights, regardless of whether they are understood as factual reports. During their impulse presentation (see agenda above), our dialogue partners used the following bible quotes to portray the Christian view on the role of humans on Earth:

"[. . .] and God saw that it was good." (Genesis 1:25)

This quote was interpreted to emphasize that nature and creation have value on their own (before humanity is created) and not only due to their potential utility for humankind.

"God blessed them; and God said to them, 'Be fruitful and multiply . . . '" (Genesis 1:28)

This sentence was said to stem from a historical context of high mortality of mothers and children. Still, "being fruitful" was seen as an important topic for the future since biodiversity—and hence fertility—is being endangered through pollution, soil erosion and climate change.

"[. . .] fill the earth, and subdue it; and rule over the fish of the sea and over the birds of the sky and over every living thing that moves on the earth." (Genesis 1:28)

In this quote, humans are explicitly given the mandate to intervene in creation and modify the environment. (Note that this interpretation of the verse is particular to the King James translation, one of the most widely used translations in the Western world, but that other translations translate this instead into the concepts of responsibility and stewardship rather than dominion; see further discussion of this below.)

"Then the LORD God took the man and put him into the Garden of Eden to cultivate it and keep it." (Genesis 2:15)

This quote was cited to portray humanity as a careful keeper of creation, capable of positive care in the sense of protecting and guarding. From these statements it was concluded that creation has a value for itself and that humans as curators and protectors of creation have received a positive mandate from God to modify and intervene in nature. For example, humans may defend themselves against negative consequences of creation (e.g., pandemics). From this perspective, humans were understood to have not only the potential to destroy but also the potential to act as careful constructor—for example, by developing cultivated land. Still, it was recognized that the whole topic of conservation of creation still represents a rather new issue in churches, an issue that is still debated internally.

Participants highlighted that the literal translation of the Hebrew wording of "subdue the Earth" (Gen 1:28) (elsewhere translated as "take dominion over the Earth") would be "to set foot on the Earth." The metaphor of "setting foot on," in turn, would be a traditional metaphor for protecting and patronizing. The mandate to humans to reign on Earth can therefore be understood as a service to creation, meaning to rule without violence, out of a mind-set of responsibility (stewardship). As an example of this suggestion that human beings can take such reasonable care for nature, a participant mentioned how the Rhine River in Germany was restored after being polluted. Another possible historical example for a kind of "geoengineering" was based upon the fact that large areas in certain German forests used to be treated with limestone in order to compensate the negative side effects of industrial flue gases and acid rain. These measures, however, were noted by the participant to have been terminated because they did not eliminate the actual root cause of the acidifying soils. Instead, industrial facilities were equipped with filters against pollutants in order to prevent the emission of harmful substances.

Participants agreed that the Bible does not set a general limitation with respect to the scale and intensity of a potential human intervention into the Earth system. However, the focus should be put on the wise weighing of advantages and disadvantages. Generally, responsibility should be understood as an arduous process, and humans should challenge themselves to undergo this hardship. The human mandate to rule and serve would mean that no part of creation should be suppressed or harmed. This requires "Solominic" wisdom, the ability to make difficult and effective decisions demonstrated by the biblical King Solomon.

AN ASSESSMENT OF CLIMATE ENGINEERING FROM A CHRISTIAN PERSPECTIVE

Against this background, participants discussed how the Christian mandate for humans to act as guardians of creation could be interpreted in the concrete context of climate engineering. They emphasized that, despite some temporary positive effects in some limited regions, climate change would lead to a negative development for humans as well as for nature. This was exemplified with a reference to Pope Benedict XVI who had stressed earlier that the climate is a good that needs to be protected.

An important basis for the discussion was the interpretation of Genesis 1:28, "subdue the earth," examining the implications of this quote with respect to climate engineering. It seemed difficult to assess whether or not the task of a steward would potentially include climate engineering. Such an assessment was considered to be only possible once the potential impacts of a specific climate engineering technology on creation were understood.

One participant summed up the general consensus that climate engineering does "not per se belong to the devil," which is to say not obviously wrong, but worthy of some suspicion. Most believed that research should be pursued, applying the principle of trial and error, but on a small scale in order to prevent undesired consequences. Indeed, it was pointed out that an integral part of human responsibility for creation includes assessing all potential options for how to act and intervene most appropriately. As mentioned above, the understanding of stewardship in a Solomonic sense would therefore require ensuring that no part of creation would be suppressed by climate engineering, meaning that no harm should be done to creation.

Given this understanding of human responsibility, a detailed understanding of all consequences and side effects would therefore be necessary before an assessment of the technologies themselves would be possible. However, many participants were skeptical about whether it would ever be possible to intervene in the Earth's complex ecosystems in a controlled manner. Also, the question of whether and how individual climate engineering measures could potentially be reversed was of central relevance. There was also significant interest in potential winners and losers from climate engineering, because it was considered essential that climate engineering should aim to increase global justice. In this context, there was concern that climate engineering could serve the interests of industrialized countries more than the interests of poor countries and therefore increase already existing injustices. Particular concern was voiced about the difficulty of establishing causal relationships between a climate engineering intervention and its impacts, which would in turn make it difficult to attribute clear responsibilities for potential consequences.

Some participants suggested that climate engineering should be understood as a potential emergency solution, particularly because they expected that several countries like India or China would not be willing to abstain from using fossil resources. From these considerations, there was concern that albedo modification technologies in particular could be deployed prematurely and unilaterally. It was therefore stressed that international governance should be in place before experimental investigations ("field tests") were to be conducted. In order to establish such a framework, however, a detailed understanding of the consequences would, again, be required.

Participants emphasized that climate engineering would be a topic embedded in a highly complex network of interests and that researchers should keep this in mind. They therefore recommended early and upfront exchanges with potential stakeholders. The ambition of the IASS to further develop a transdisciplinary approach to climate engineering research and dialogue was therefore considered helpful, and was explicitly supported.

Very fundamentally, participants also argued that the history of technological development does not represent a story of sustainable problem solving. The hope and expectation that technology could ultimately solve the problems of our time "should therefore actually be over now," one of the participants said.

CONCRETE RECOMMENDATIONS FOR RESEARCH ON CLIMATE ENGINEERING

Participants did not see the church's mandate as making concrete technological suggestions or assessing specific technologies proposed by others. The church's mandate was rather considered as a contribution to an "Ethical Engineering," as for example reflected in the statement "If our attitude does not change, all technology will not help." Acknowledging the importance of the principle of freedom of research, participants voiced the opinion that the tendency of religious groups to make statements in the tone of "you should" and "you should not" is not always helpful.

It was recognized that further research in the field of climate engineering is needed in order to ensure that Germany or also the European Union would be able to take a position based on deep understanding of the issues involved. It seemed important in this context that the EU aims mainly for a better understanding of climate engineering, not in order to promote a potential deployment, but rather to better influence the political development in this field together with the international community. Participants hoped that political leaders would be equipped to take a better position against stakeholders that might propagate a premature deployment of climate engineering measures.

Participants recommended that climate engineering researchers clearly distinguish between carbon dioxide removal and albedo modification. Research on carbon dioxide removal was recommended to include investigations on the consequences as well as on the possibilities of deployment, whereas for albedo modification it was recommended to only focus on the investigation of consequences and governance. Generally, they agreed that research on climate change mitigation and possible adaptation measures should be prioritized in the future.

Another recommendation was the inclusion of an historian in discussions of climate engineering, in order to assess historic examples of human interventions in nature and to identify principles when and why such interventions caused undesired consequences. From the perspective of Christian ethics, the problem of inequality with respect to winners and losers appears to be central. It was recommended that the questions "Who would benefit from climate engineering?" and "Who would suffer from climate engineering?" should be examined, considering particularly the consequences of climate engineering on the basic needs of the poorest. In this context, participants emphasized that the Brundtland Report has already attributed highest priority to the basic needs of the poor. This aspect should be addressed particularly by research on climate engineering. Model simulations and consequent assessments were, for example, suggested to assess not only regional disparities of potential impacts, but particularly the impacts on different social groups.

These observations agree well with the conclusions described by Corner et al. about the stakeholder engagement on climate engineering in the UK.[24] While they observed that the general knowledge about climate engineering among members of the public is low, they found generally more public support for CDR than SRM and a preference of conventional climate change mitigation policies over climate engineering. Of the potentially emerging themes identified by Corner et al. for further deliberative research the participants of the dialogues particularly raised the capacity for responsible and transparent governance. This suggests that our experiences and findings may not be specific to the engagement with German Catholic groups but that our conclusions may hold also outside the German context.

LESSONS LEARNED WITH RESPECT TO TRANSDISCIPLINARITY AND COMMUNICATION WITH RELIGIOUS GROUPS

As mentioned earlier, the IASS dialogue series on climate engineering was established in early 2013 when the debate on climate engineering was still at a relatively early stage in Germany. Most potential stakeholders in Germany were hardly aware of the topic, and for many groups, including religious groups, the IASS's initial approach was often the first time that they heard of climate engineering.

Given this starting point, it turned out to be unrealistic to advance directly to the stage of forming a common research object as described in phase 1 ("problem transformation") of the model proposed by Jahn et al.[25] Instead, the meetings had an explorative character, raising awareness among potential stakeholders about the relevance of climate engineering and in turn developing sensitivity among researchers on issues that could potentially be societally relevant. While the exchange did actually lead to several valuable (partly joint) activities afterward and helped to establish trustful contacts, no immediate problem was identified for which new knowledge needed to be developed jointly. We therefore consider our dialogue series to represent only the beginning of a transdisciplinary approach to climate engineering.

Since the establishment of the dialogue series, a lot has changed in the debate about climate engineering in Germany, raising the question of how to continue or further develop the approach of the Climate Engineering Research Microcosm in this context. The awareness of the potential stakeholders for the topic has increased, and much of the German mass media has reported on the topic. Carbon Dioxide Removal has gained particular attention and acceptance, boosted also by the fifth assessment report of the IPCC, which mentioned climate engineering prominently and included specific CDR technologies in its mitigation scenarios for keeping the 2°C limit.[26]

Against this background, we see the situation changing in a way that might allow us to advance a step further in our ambition to develop a transdisciplinary approach on climate engineering. More stakeholders are beginning to see the topic as potentially relevant and might therefore be more willing to engage in a joint process of solving a specific societal problem together with academics. Exploring this "societal problem space" better in order to identify societal partners for a concrete joint problem solving process could be one perspective for the further development of the dialogue series.

While we do see our dialogue series only as first steps toward developing a transdisciplinary approach on climate engineering, the exchange still exhibited several characteristics that previous authors have identified as typical for transdisciplinary approaches and that we consider an essential basis for our further progress. We found our format to be very suitable to facilitate genuine dialogue, and by framing the meetings as opportunities to listen rather than lecture, we tried to apply the mind-set of humility and a willingness to contextualize our own knowledge as described by Lawhon et al.[27] This seemed to be appreciated by our exchange partners.

The need for humility is particularly important for the exchange with religious perspectives. Naturally, the languages and perspectives of people with religious or spiritual backgrounds deviated significantly from the ones of the involved academic researchers (which originated from all kinds of disciplines but not from theology or researchers in religion) and in some first exchange settings, some participants were alienated from these differences. Overcoming this alienation and engaging genuinely with religious representatives was certainly a particular challenge, and for the future, scholars with backgrounds in theology or religious studies could become important for building bridges as they are familiar with both the academic culture as well as the religious perspectives. Also against this background, it was meanwhile decided to expand the Climate Engineering Research Microcosm by adding a scholar with degrees in both theology and biology, who is going to explicitly address the connections between climate engineering and religion and engage in the respective dialogue processes.

Generally, we found it helpful when participants were explicit and transparent about the assumptions and values underlying their opinions and tried to avoid persuading each other or judging each other's perspectives, as, for example, also described by Wynne.[28] A mind-set of nonsuperiority (of science over religion as well as the other way around) helped significantly in facilitating the genuinely dialogical culture that is essential in transdisciplinary processes, as discussed by several authors.[29] However, genuinely valuing each other's perspectives and being willing to contextualize assumptions and self-conceptions seemed challenging for the individuals involved in such processes. We would therefore encourage ongoing, deeper research on how transdisciplinary processes are influenced by the mind-sets of the involved individuals and how certain mind-sets that are helpful for these processes can be fostered and cultivated methodologically.

SUMMARY AND OUTLOOK

In this chapter, we have described our experiences of exchange with nonacademic stakeholders, especially with religious groups on the topic of climate engineering. These experiences represent first steps of the Climate Engineering Research Microcosm of the IASS toward developing a transdisciplinary approach to research and dialogue on climate engineering. Throughout this process, we have learned to understand some of the aspects of climate engineering that appear particularly critical from a Christian point of view, and we have learned to better understand the specific language that is relevant in this context.

We have also learned that from a Christian perspective it is not completely clear how to assess the limits of human interventions in nature. The Bible seems to set no principle limit to the scale at which humans have the mandate to intervene in creation. It seemed essential, instead, that the Bible depicts human responsibility toward creation as an arduous process. People should aim to reduce harm to all beings—that is, human dominion over the Earth should be understood as ruling in a Solomonic sense. From a Christian perspective, investigations of climate engineering techniques should undergo this "arduous process of responsibility" and would therefore need to assess the potential impacts on creation as holistically as possible before an implementation is considered. In this context, a particular emphasis was placed on assessing the potential impacts of climate engineering on the poorest people on Earth. Not only regional disparities, but also social disparities of impact seemed essential to the participants.

NOTES

1. The authors would like to thank the organizers and participants of the track *The Technological Response: Geo-Engineering* at the conference *Seizing an Alternative: Toward an Ecological Civilization* for the enriching discussion and helpful comments, Falk Schmidt for further discussion about transdisciplinarity, and the book editors for their constructive criticism of the chapter.

2. David W. Keith, "The Earth Is Not Yet an Artifact," *Technology and Society Magazine, IEEE* 19, no. 4 (2000): 25–28; David W. Keith, "Geoengineering the Climate: History and Prospect," *Annual Review of Energy and the Environment* 25 (2000): 245–84; Stephen H. Schneider, "Earth Systems Engineering and Management," *Nature* 409, no. 6818 (2001): 417–21; Paul J. Crutzen, "Geology of Mankind," *Nature* 415, no. 6867 (2002): 23; Paul J. Crutzen, "Albedo Enhancement by Stratospheric Sulfur Injections: A Contribution to Resolve a Policy Dilemma?" *Climatic Change* 77, no. 3–4 (2006): 211–20.

3. Marcia McNutt et al., *Climate Intervention: Reflecting Sunlight to Cool Earth* (Washington DC: National Academy of Sciences, 2015); Marcia McNutt et al., *Climate Intervention: Carbon Dioxide Removal and Reliable Sequestration* (Washington DC: National Academy of Sciences, 2015); S. Schäfer et al., "The European Transdisciplinary Assessment of Climate Engineering (EuTRACE): Removing Greenhouse Gases from the Atmosphere and Reflecting Sunlight Away from Earth," 2015.

4. Forrest Clingerman, "Geoengineering, Theology, and the Meaning of Being Human," *Zygon* 49, no. 1 (2014): 6–21; Adam Corner, Nick Pidgeon, and Karen Parkhill, "Perceptions of Geoengineering: Public Attitudes, Stakeholder Perspectives, and the Challenge of 'Upstream' Engagement," *Wiley Interdisciplinary Reviews: Climate Change* 3, no. 5 (2012): 451–66.

5. For example, in the titles of the contributions by James R. Fleming, "The Climate Engineers: Playing God to Save the Planet," *The Wilson Quarterly* (Spring 2007): 46–60; and Forrest Clingerman and Kevin J. O'Brien, "Playing God: Why Religion Belongs in the Climate Engineering Debate," *Bulletin of the Atomic Scientists* (2014): 27–37.

6. Will Steffen, Åsa Persson, Lisa Deutsch, Jan Zalasiewicz, Mark Williams, Katherine Richardson, Carole Crumley, et al., "The Anthropocene: From Global Change to Planetary Stewardship," *Ambio* 40, no. 7 (2011): 739–61; Will Steffen, Paul J Crutzen, and John R

McNeill, "The Anthropocene: Are Humans Now Overwhelming the Great Forces of Nature," *Ambio: A Journal of the Human Environment* 36, no. 8 (2007): 614–21.

7. Thomas Jahn, Matthias Bergmann, and Florian Keil, "Transdisciplinarity: Between Mainstreaming and Marginalization," *Ecological Economics* 79 (2012): 1–10.

8. Julie Thompson Klein, "Prospects for Transdisciplinarity," *Futures* 36, no. 4 (2004): 515–26; Gertrude Hirsch Hadorn, David Bradley, Christian Pohl, Stephan Rist, and Urs Wiesmann, "Implications of Transdisciplinarity for Sustainability Research," *Ecological Economics* 60, no. 1 (2006): 119–28; Carlo Jaeger, J David Tàbara, and Julia Jaeger, *European Research on Sustainable Development: Volume 1: Transformative Science Approaches for Sustainability* (Berlin, Heidelberg: Springer-Verlag, 2011). See also Jerome R. Ravetz, "Post-Normal Science and the Complexity of Transitions towards Sustainability," *Ecological Complexity* 3, no. 4 (2006): 275–84.

9. Jahn et al., "Transdisciplinarity."

10. Some authors understand transdisciplinarity as a challenge to the primacy of rational thought as cultivated in the Western scientific system. Beyond seeing transdisciplinarity merely as a practical approach for problem solving in a more systemic way in order to deal with complexity (*weak transdisciplinarity*), they consider transdisciplinarity an open project that aims toward a different manner of seeing the world more holistically, embracing different levels of reality and intuitive *understanding* rather than only *knowledge* (*strong transdisciplinarity*). For further details on this see, for example, the discussion in Jahn, "Transdisciplinarity:," Manfred A. Max-Neef, "Foundations of Transdisciplinarity," *Ecological Economics* 53, no. 1 (2005): 5–16; Alfonso Montuori, "Complexity and Transdisciplinarity: Reflections on Theory and Practice," *World Futures: The Journal of New Paradigm Research* 69, no. 4–6 (2013): 200–30; Edgar Morin, "From the Concept of System to the Paradigm of Complexity," *Journal of Social and Evolutionary Systems* 15, no. 4 (1992): 371–85; Edgar Morin, *On Complexity* (Hampton Press, 2008); Basarab Nicolescu, *La Transdisciplinarité: Manifeste* (English Translation by K-C. Voss, *Manifesto of Transdisciplinarity* [Albany: SUNY Press, 2001]) (Paris: Editions du Rocher, 1996); Ervin Laszlo, *The Connectivity Hypothesis: Foundations of an Integral Science of Quantum, Cosmos, Life, and Consciousness* (Albany: SUNY Press, 2010).

11. Uwe Schneidewind, "Mobilizing the Intellectual Capital of Universities," in *Transdisciplinarity: Joint Problem-Solving among Science, Technology and Society*, edited by J. Thompson Klein et al., 94–100 (Basel: Birkhauser, 2001).

12. Jahn et al., "Transdisciplinarity"; D. J. Lang, A. Wiek, M. Bergmann, M. Stauffacher, P. Martens, P. Moll, M. Swilling, and C. J. Thomas, "Transdisciplinary Research in Sustainability Science: Practice, Principles, and Challenges," *Sustainability Science* 7, SUPPL. 1 (2012): 25–43.

13. Jahn et al., "Transdisciplinarity," Nicolescu, *La Transdisciplinarité.*

14. Florin Popa, Mathieu Guillermin, and Tom Dedeurwaerdere, "A Pragmatist Approach to Transdisciplinarity in Sustainability Research: From Complex Systems Theory to Reflexive Science," *Futures* 65 (2015): 45–56.

15. Thompson Klein, "Prospects for Transdisciplinarity," 515–26.

16. Mary Lawhon, Panate Manomaivibool, and Hiromi Inagaki, "Solving/Understanding/ Evaluating the E-Waste Challenge through Transdisciplinarity?" *Futures* 42, no. 10 (2010): 1212–21.

17. Jahn et al., "Transdisciplinarity."

18. K. Töpfer, "Wissenschaftliche Arbeiten Des Institute for Advanced Sustainability Studies," (2010).

19. B. Wynne, "Public Engagement as a Means of Restoring Public Trust in Science: Hitting the Notes but Missing the Music," *Community Genetics* 9 (2006): 211–30.

20. Wylie A. Carr, Christopher J. Preston, Laurie Yung, Bronislaw Szerszynski, David W. Keith, and Ashley M. Mercer, "Public Engagement on Solar Radiation Management and Why It Needs to Happen Now," *Climatic Change* 121, no. 3 (2013): 567–77.

21. For social scientific research on stakeholder engagement about climate engineering see, for example, Corner et. al., "Perceptions of Geoengineering"; Phil Macnaghten and Bronislaw Szerszynski, "Living the Global Social Experiment: An Analysis of Public Discourse on Solar

Radiation Management and Its Implications for Governance," *Global Environmental Change* 23, no. 2 (2013): 465–74; Malcolm J. Wright, Damon A. H. Teagle, and Pamela M. Feetham, "A Quantitative Evaluation of the Public Response to Climate Engineering," *Nature Climate Change* 4, no. 2 (2014): 106–10.

22. Klein, "Prospects for Transdisciplinarity."

23. Corner et al., "Perceptions of Geoengineering"; Macnaghten and zerszynski, "Living the Global Social Experiment"; Wright et al., "A Quantitative Evaluation of the Public Response to Climate Engineering."

24. Corner et al.,"Perceptions of Geoengineering."

25. Jahn et al., "Transdisciplinarity."

26. IPCC, "Climate Change 2014—Mitigation of Climate Change. Contribution of Working Group III to the Fifth Assessment Report of the Intergovernmental Panel on Climate Change" (Cambridge, United Kingdom,and New York, USA, 2014).

27. Lawhon et al., "Solving/Understanding/Evaluating the E-Waste Challenge through Transdisciplinarity?"

28. Wynne, "Public Engagement as a Means of Restoring Public Trust in Science."

29. Jahn et al., "Transdisciplinarity"; Thompson Klein, "Prospects for Transdisciplinarity"; Popa, "A Pragmatist Approach to Transdisciplinarity in Sustainability Research"; Lawhon et. al., "Solving/Understanding/Evaluating the E-Waste Challenge through Transdisciplinarity?"; Wynne, "Public Engagement as a Means of Restoring Public Trust in Science"; Carr et al,. "Public Engagement on Solar Radiation Management and Why It Needs to Happen Now."

BIBLIOGRAPHY

Carr, Wylie A., Christopher J. Preston, Laurie Yung, Bronislaw Szerszynski, David W. Keith, and Ashley M. Mercer. "Public Engagement on Solar Radiation Management and Why It Needs to Happen Now." [In English]. *Climatic Change* 121, no. 3 (2013): 567–77.

Clingerman, Forrest. "Geoengineering, Theology, and the Meaning of Being Human." *Zygon* 49, no. 1 (2014): 6–21.

Clingerman, Forrest, and Kevin J. O'Brien. "Playing God: Why Religion Belongs in the Climate Engineering Debate." *Bulletin of the Atomic Scientists* (2014): 27–37.

Corner, Adam, Nick Pidgeon, and Karen Parkhill. "Perceptions of Geoengineering: Public Attitudes, Stakeholder Perspectives, and the Challenge of 'Upstream' Engagement." *Wiley Interdisciplinary Reviews: Climate Change* 3, no. 5 (2012): 451–66.

Crutzen, Paul J. "Geology of Mankind." *Nature* 415, no. 6867 (2002): 23.

Crutzen, Paul J. "Albedo Enhancement by Stratospheric Sulfur Injections: A Contribution to Resolve a Policy Dilemma?" *Climatic Change* 77, no. 3–4 (2006): 211–20.

Fleming, James R. "The Climate Engineers: Playing God to Save the Planet." *The Wilson Quarterly* (Spring 2007): 46–60.

Gibbons, Michael, Camille Limoges, Helga Nowotny, Simon Schwartzman, Peter Scott, and Martin Trow. *The New Production of Knowledge: The Dynamics of Science and Research in Contemporary Societies*. Newbury Park, CA, London: Sage, 1994.

Hirsch Hadorn, Gertrude, David Bradley, Christian Pohl, Stephan Rist, and Urs Wiesmann. "Implications of Transdisciplinarity for Sustainability Research." *Ecological Economics* 60, no. 1 (2006): 119–28.

IPCC. "Climate Change 2014—Mitigation of Climate Change. Contribution of Working Group III to the Fifth Assessment Report of the Intergovernmental Panel on Climate Change." Cambridge, United Kingdom, and New York, USA, 2014.

Jaeger, Carlo, J David Tàbara, and Julia Jaeger. *European Research on Sustainable Development: Volume 1: Transformative Science Approaches for Sustainability*. Berlin, Heidelberg: Springer-Verlag, 2011.

Jahn, Thomas, Matthias Bergmann, and Florian Keil. "Transdisciplinarity: Between Mainstreaming and Marginalization." *Ecological Economics* 79, (2012): 1–10.

Keith, D. W. "The Earth Is Not Yet an Artifact." *Technology and Society Magazine, IEEE* 19, no. 4 (2000): 25–28.

Keith, David W. "Geoengineering the Climate: History and Prospect." *Annual Review of Energy and the Environment* 25 (2000): 245–84.

Lang, D. J., A. Wiek, M. Bergmann, M. Stauffacher, P. Martens, P. Moll, M. Swilling, and C. J. Thomas. "Transdisciplinary Research in Sustainability Science: Practice, Principles, and Challenges." *Sustainability Science* 7, SUPPL. 1 (2012): 25–43.

Laszlo, Ervin. *The Connectivity Hypothesis: Foundations of an Integral Science of Quantum, Cosmos, Life, and Consciousness*. SUNY Press, 2010.

Lawhon, Mary, Panate Manomaivibool, and Hiromi Inagaki. "Solving/Understanding/Evaluating the E-Waste Challenge through Transdisciplinarity?" *Futures* 42, no. 10 (2010): 1212–21.

Macnaghten, Phil, and Bronislaw Szerszynski. "Living the Global Social Experiment: An Analysis of Public Discourse on Solar Radiation Management and Its Implications for Governance." *Global Environmental Change* 23, no. 2 (2013): 465–74.

Max-Neef, Manfred A. "Foundations of Transdisciplinarity." *Ecological Economics* 53, no. 1 (2005): 5–16.

McNutt, M., et al. *Climate Intervention: Carbon Dioxide Removal and Reliable Sequestration*. Washington D.C. : National Academy of Sciences, 2015.

———. *Climate Intervention: Reflecting Sunlight to Cool Earth*. Washington D.C.: National Academy of Sciences, 2015.

Montuori, Alfonso. "Complexity and Transdisciplinarity: Reflections on Theory and Practice." *World Futures: The Journal of New Paradigm Research* 69, no. 4–6 (2013): 200–30.

Morin, Edgar. "From the Concept of System to the Paradigm of Complexity." *Journal of Social and Evolutionary Systems* 15, no. 4 (1992): 371–85.

———. *On Complexity*. Hampton Press, 2008.

Nicolescu, Basarab. *La Transdisciplinarité: Manifeste* (English Translated by K-C. Voss, *Manifesto of Transdisciplinarity*. State University of New York Press, 2001). Paris: Editions du Rocher, 1996.

———. *Transdisciplinarity: Theory and Practice*. Hampton Pr, 2008.

Popa, Florin, Mathieu Guillermin, and Tom Dedeurwaerdere. "A Pragmatist Approach to Transdisciplinarity in Sustainability Research: From Complex Systems Theory to Reflexive Science." *Futures* 65, (2015): 45–56.

Ravetz, Jerome R. "Post-Normal Science and the Complexity of Transitions towards Sustainability." *Ecological Complexity* 3, no. 4 (2006): 275–84.

Schäfer, S., et al. "The European Transdisciplinary Assessment of Climate Engineering (EuTRACE): Removing Greenhouse Gases from the Atmosphere and Reflecting Sunlight Away from Earth." 2015.

Schneider, Stephen H. "Earth Systems Engineering and Management." *Nature* 409, no. 6818 (2001): 417–21.

Schneidewind, Uwe. "Mobilizing the Intellectual Capital of Universities." In *Transdisciplinarity: Joint Problem-Solving among Science, Technology and Society*, edited by J. Thompson Klein et al., 94–100. Basel: Birkhauser, 2001.

Steffen, Will, Åsa Persson, Lisa Deutsch, Jan Zalasiewicz, Mark Williams, Katherine Richardson, Carole Crumley, et al. "The Anthropocene: From Global Change to Planetary Stewardship." *Ambio* 40, no. 7 (2011): 739–61.

Steffen, Will, Paul J. Crutzen, and John R. McNeill. "The Anthropocene: Are Humans Now Overwhelming the Great Forces of Nature." *Ambio: A Journal of the Human Environment* 36, no. 8 (2007): 614–21.

Thompson Klein, Julie. "Prospects for Transdisciplinarity." *Futures* 36, no. 4 (2004): 515–26.

Töpfer, K. "Wissenschaftliche Arbeiten Des Institute for Advanced Sustainability Studies." (2010).

Wright, Malcolm J., Damon A. H. Teagle, and Pamela M. Feetham. "A Quantitative Evaluation of the Public Response to Climate Engineering." *Nature Climate Change* 4, no. 2 (2014): 106–10.

Wynne, B. "Public Engagement as a Means of Restoring Public Trust in Science: Hitting the Notes but Missing the Music." *Community Genetics* 9 (2006): 211–30.

II

Philosophical and Theological Responses to Climate Engineering

Chapter Three

The Temptations of Climate Engineering

Dane Scott

There is increasing anxiety that the world's great economies will not act in time to avoid dangerous climate change. Many fear that the lack of action in reducing greenhouse gas emissions will lead to much human suffering and tremendous loss of the earth's biodiversity. In response, some scientists and policymakers are now taking climate engineering—a technological fix to slow or stall the effects of climate change—seriously. The most discussed class of climate engineering schemes is solar radiation management (SRM). SRM involves manipulating the earth's solar budget by either dispersing reflective aerosols in the stratosphere or increasing the reflective capacity of the earth's surface. The focus of this chapter will be on SRM techniques involving stratospheric aerosols; however, much of what is said might be applied to other climate engineering schemes.

One common reaction to SRM is to worry that engineering the earth's climate could lead to serious environmental disasters. These fears are no doubt in part driven by recent experiences, such as the massive oil spill in the Gulf of Mexico from the Deepwater Horizon incident and the Fukushima Daiichi nuclear power catastrophe on the coast of Japan. In both events, people around the world helplessly watched live video of crude oil and radioactive pollution being relentlessly discharged into the environment.

There are multiple causes for technological disasters like the Deepwater Horizon and Fukushima Daiichi, including "acts of God" (events out of human control, such as earthquakes), engineering errors, manufacturing errors, regulatory failures, and cultural obstacles to right actions. However, such technological tragedies are at least in part due to failures of human conduct and character, created not only by the actions of individuals, but also by complex interactions between political, economic, and technological systems and human character and cultures.

FOIBLES OR SIN

In a paper I coauthored with Jane C. S. Long, discussing the problem of vested interest and climate engineering governance, we began by noting that the Fukushima Daiichi nuclear disaster was in part the result of people and institutions protecting vested interests.[1] That paper focused more on character and culture than political, economic, and technological systems. As we moved from the initial example of Fukushima to climate engineering, we identified four types of vested interest that could undermine fundamental public trust in climate engineering research and possible deployment. These were labeled the 4 Fs: fortune, fear, fame, and fanaticism. Long and I observed: "Society may, at some future time, attempt geoengineering in order to stave off the worst, most unbearable effects of climate change. The technical challenge alone is enormous, but Fukushima provides a cautionary tale for managing the endeavor. Is it possible to develop the trustworthy capacity to manage the climate of Earth?"[2] In the first half of that paper, we identified potential problems in climate engineering research and governance caused by human foibles and vested interests. In the second half, we made specific recommendations for building a research and governance culture that would mitigate some of the negative consequences of those problems.

Since our focus was on human foibles and vested interest, early in the process of collaborating on that paper I pondered whether "foible" was the best term for our analysis. Foible is perhaps only a quasi-ethical term, pointing to minor weaknesses or flaws of character. When one looks at the magnitude of disasters like Deepwater Horizon or Fukushima, a stronger term is needed to describe the moral failures that contributed to these catastrophes. However, I concluded that for the audience and purposes of that paper, foible was a good choice. I am not sure how the readers of a journal devoted to issue in science and technology would have reacted to the alternative theological terms I have chosen for the analysis in this chapter: temptation and sin.

Given the theme of this volume—climate engineering, theology, and religious ethics—readers might find it more acceptable, if barely, to focus on these terms. The religious notion of sin will appear to many people as unpopular and outmoded. Nonetheless, the influential twentieth-century Christian ethicist Reinhold Niebuhr offers an analysis of temptation and sin that provides insights into the threats that vested interests pose to trustworthy climate engineering research, governance, and possible deployment.

Niebuhr's discussions of temptation and sin are not narrowly theological. They have broad implications for ethics and public affairs. Further, his insights into politics and international relations gained him the admiration of many atheistic or agnostic political theorists. In fact, he has been criticized for secularizing theological doctrines. But according to Richard Crouter, that criticism misses the point: "It is not so much that Niebuhr secularized but that he humanized the age-old Christian teaching."[3] By focusing on phenomenological and pragmatic aspects of the notions of temptation and sin, Niebuhr provides an insightful set of related concepts for understanding human character and the behaviors of nations. Niebuhr was a public theologian, and his ethics have proven useful in many public policy arenas. In particular, I will argue that Niebuhr's insights can be applied to climate engineering politics.

MORAL HAZARD OR TEMPTATION

In current discussions about climate engineering, for reasons that are not entirely clear, "moral hazard" has become a blanket term that would likely cover issues like the "temptations" that arise from SRM. The moral hazard argument is a warning against possible unintended consequences of climate engineering arising from its effect on human behaviors. The term is borrowed from the insurance industry, where it identifies an unwanted side effect of products like work hazard insurance. The general idea is that work hazard insurance will lead to more accidents by increasing risky behaviors. The moral hazard argument for climate engineering asserts that climate engineering would somehow intentionally promote the risky behaviors that caused the problems it is trying to fix. The appropriateness and utility of this term as applied to climate engineering has been challenged.[4] I will not discuss those challenges here. However, I hope that the remainder of this chapter will demonstrate that Niebuhr's analysis of temptation and sin will be more illumining than the concept of moral hazard. Toward that goal, since the notions of temptation and sin are likely to seem outdated to many, it is important to explain why it is worth returning to Niebuhr.

WHY NIEBUHR NOW?

Over the last decade there has been renewed interest in the thought of Reinhold Niebuhr. Richard Crouter's explanation for this phenomenon is that "Niebuhr's account of 'the grandeur and misery of man' (Pascal's phrase) has always received more attention during times of crisis."[5] The Iraq War, the War in Afghanistan and the War on Terror have created a time of crisis, and several important thinkers have mined Niebuhr's writings for insights to critically consider these conflicts.

Prior to the United States entering World War II, Niebuhr, a former pacifist, was an influential voice calling for the United States to enter the war to confront the evil of Nazism. However, during the Cold War Niebuhr served as a voice for America's moral conscience by consistently warning that the United States' military power combined with its moral idealism was a setup for "sin." In the *Irony of American History*, Niebuhr writes:

> We may be too secure in both our sense of power and our sense of virtue to be ready to engage in a patient chess game with the recalcitrant force of historic destiny. We could bring calamity upon ourselves and the world by forgetting that even the most powerful nations and even the wisest planners of the future remain themselves creatures as well as creators of the historical process.[6]

Interestingly, critics of the United States' military policies after September 11th (e.g., Andrew Bacevich, *The Limits of Power*) and defenders (e.g., Jean Bethke Elshtain, *Just War against Terror*) have appealed to different aspects of Niebuhr's thought to justify their positions. Elshtain appealed to the need to confront evil with force, and Bacevich appealed to the dangers of moral overreach in the exercise of force. This does not mean that Niebuhr was inconsistent; he provided different analyses and responses to different circumstances. Elshtain and Bacevich held different interpretations of the circumstances, which lead them to emphasize different aspects of Niebuhr's thoughts.

Another reason for renewed interest in Reinhold Niebuhr has to do with his influence on President Obama. In a 2007 interview, then-candidate Senator Obama identified Niebuhr as his favorite theologian. President Obama demonstrated a depth of engagement with Niebuhr's writings as he adroitly summarized central themes. The President notes:

> I take away [from Niebuhr] the compelling idea that there is serious evil in the world, and hardship and pain. And we should be humble and modest in our belief that we can eliminate these things. But we shouldn't use that as an excuse for cynicism and inaction. I take away . . . the sense we have to make these efforts knowing they are hard, and not swing from naïve idealism to bitter realism.[7]

The central themes in Niebuhr's thought that President Obama summarized can be readily applied to global climate change politics. For instance, the pain and suffering that is and will be caused by a rapidly changing climate must be seen as a serious evil. Further, it will not be possible to completely eliminate these evils either through mitigation or adaptation policies, or climate engineering. Human efforts will inevitably be limited and flawed, like their authors. Nevertheless, the impurity and imperfection of human action is no excuse for cynicism or idleness. The extraordinarily hard task of dealing with climate change must be done with modesty and humility, but also with courage and resolve. Finally, one must equally avoid the naïve idealism of overwrought moralistic crusades to save the planet and the bitter realism of believing that nations are only capable of acting in their narrow self-interest.

In a recent essay that connects Niebuhr's thought with contemporary challenges, Anatol Lieven begins by noting that "climate change looms as a unprecedented danger to modern civilization as a whole."[8] He observed, further, that acting on climate change requires nations to move beyond short term self-interest and participate in an unprecedented amount of cooperation. Along these lines, Niebuhr wrote: "A consistent self-interest on the part of a nation will work against its interest because it will fail to do justice to its broader and longer interests, which are involved with those of other nations."[9] Powerful countries and political unions like the United States and European Union cannot use their power to act unilaterally on reducing greenhouse gases. All the major developed and developing countries—the European Union, United States, China, India, Russia, Brazil, etc.—must to a significant degree compromise and cooperate to reduce threats of dangerous climate change.

Lieven thinks that Niebuhr's Christian realism avoids cynicism that acknowledges only political power, and allows for the possibility of morally driven change. This sober, self-critical optimism can serve as a guide to dealing with a contemporary problem like climate change by merging key insights from "the realist and progressive traditions."[10] Niebuhr's Christian realism understands humans as fallen and imperfect, but we nevertheless have a duty to "seek the good insofar as we are able."[11] Jean Bethke Elshtain summarizes Niebuhr's position similarly, writing that "we can and must work toward a world less violent, horrific, and cruel. But we cannot definitively cure the universe. Those who attempt such cures wind up with wild moralistic crusades that trail even further horror in their wake."[12] Nowhere is the need to avoid the extremes of idealistic moralistic crusades and defeatist cynical realism and to embrace a hopeful realism that works toward cooperation more evident than in climate change politics. Niebuhr's Christian realism is optimistic enough to encourage sustained efforts at building cooperation but realistic enough to know that cooperation will always be imperfect and tainted by self-interest.

MITIGATION POLICIES AND CLIMATE ENGINEERING

It is key for this discussion on the temptations of climate engineering to consider the commonsense relationship between compromise, cooperation, mitigation policies, and SRM. International climate change politics and the prospects for building cooperation are now inevitably altered, to at least some extent, by the possible option of climate engineering. Once climate engineering proposals began to be seriously discussed, they changed the policy equation by adding another set of possible options alongside the existing sets of mitigation and adaptation policies and technologies. A simplistic but instructive way to characterize the relationship is that the more successful climate change mitigation and adaptation policies are, the less interest there will be in discussing climate engineering. But the converse is also true: the less successful mitigation and adaptation policies are, the more climate engineering will move to the fore. Since the success of mitigation and adaptation policies are a function of international cooperation, the less major countries compromise and cooperate, the more seriously climate engineering options will be considered.

One reason for this is that SRM technologies do not necessarily require the same extent of cooperation between nations as mitigation polices. A cohort of nations, a single nation, or even a wealthy individual could, in principle, unilaterally launch an SRM program. Climate engineering experts have somewhat famously noted that "SRM has three essential characteristics: it is cheap, fast and imperfect."[13] The qualities of fast and cheap mean that disbursing reflective sulfate particles into the upper atmosphere can, from an engineering perspective, be accomplished without much international cooperation. The qualities of imperfect and uncertain mean that SRM is not a substitute for reducing greenhouse gas emissions and its effects cannot be completely predicted. Ricke et al. write:

> Unlike the control of greenhouse gas emissions, which must be undertaken by all major emitting nations to be effective and is likely to be costly, geoengineering could be undertaken quickly and unilaterally by a single party, at relatively low cost. Unilateral geoengineering, however, is highly likely to impose costs on other countries and run risks with the entire planet's climate system.[14]

While the effects of an SRM program on lowering the earth's average temperature can likely be predicted with some accuracy, its consequences on finer grained, regional weather patterns are far more uncertain. The relatively predictable rainfall patterns in many places that are essential for crops, water supplies, forests, and grasslands could be affected in unexpected ways by an SRM program. For example, Alan Robock et al. conducted experiments using climate models to predict the effects of attempting to cool the Arctic with SO_2. They found that one possible unintended consequence of these efforts could be to disrupt the rainfall patterns of the Asian and African monsoons on which billions of people rely for food.[15] SRM is imperfect and uncertain, and so its use includes the serious threat of environmental disasters. The fact that this technology is, relatively, fast and cheap creates a temptation to sin.

THE TEMPTATIONS OF SRM

There are least two general frames and implied arguments commonly used to justify research and possible deployment of SRM: Plan B and Insurance Policy.[16] "The Plan B" frame implies that climate engineering might be capable of buying back some of the time lost due to decades of political inaction on international mitigation and adaptation efforts. The "Insurance Policy" argument frame suggests that we may need a ready response to the potential threats of climate emergencies that result from human-caused, non-linear climate change. While these framings and arguments are controversial, they raise serious issues and real questions. I will not review the merits, or lack of merits, of these frames and arguments here. What is important for this discussion is how they demonstrate, if you will, that SRM presents a temptation to the sin of overreach, or hubris. This is not to say that scientists and decision makers should shun SRM research, but that they should understand these temptations and take actions to guard against them in climate engineering politics.

Again, the Plan B argument indicates that climate engineering might be needed to buy more time for mitigation and adaptation policies to be implemented and become effective. In other words, the window of opportunity for political actions to avoid some of the worst consequences of climatic change is quickly closing. For example, a commonly stated goal of mitigation policies is to reduce GHG emissions to a level capable of keeping the rise in the earth's average global temperature to no more than 2 degrees Celsius over the next century. One might label this "Plan A." However, it is increasingly likely that international climate negotiations will not achieve this goal and we will need a Plan B. While the 2015 United Nation's Climate talks in Paris have been heralded as a positive step forward and a "triumph of international diplomacy," the nonbinding pledges made by participating countries still fall short of the 2 degrees Celsius target, let alone the goal of 1.5 Celsius that was increasingly discussed at the talks.[17] In fact, the latest United Nations Intergovernmental Panel on Climate Change report indicates that it is increasingly unlikely to stay below the 2 degrees Celsius threshold.[18] This all but guarantees more serious discussions on the need for a Plan B. The famous climate scientist James Hansen is quoted as saying that "[the Paris agreement is] worthless words. There is no action, just promises. As long as fossil fuels appear to be the cheapest fuels out there, they will continue to be burned."[19] While the Paris conference has generated hope, it has done little to diminish the possible need for a plan B. This would be climate engineering, and the leading candidates at moment are SRM proposals involving dispersion of reflective aerosols (most likely sulfate particles) into the stratosphere. Again, the idea is that SRM can keep the earth's average global temperature below dangerous thresholds while mitigation and adaptation efforts continue with greater earnestness and effectiveness. Once these efforts meet certain emission targets and adaptation goals, then climate engineering could be wound down. The late climate scientist Stephen Schneider likened this approach to using methadone to help treat heroin addiction. SRM would be a crutch while the world's economies overcome their addiction to fossil fuels.[20]

The Insurance Policy argument focuses on the possibility of needing a ready response to a climate emergency. The fear is that there might be tipping points that would cause the earth's climate to change at a dangerously rapid pace. There are at least three candidates for climate emergencies that would trigger such tipping points: "irreversible melt of the Greenland ice sheet, massive die back of the Amazon rainforest, and a shift in the West African monsoon."[21] A climate emergency would have serious consequences for civilizations and biodiversity. However, early warnings might be possible, so in theory scientists and decision makers would have time to react to a climate emergency with an SRM project. Of course, the science of climate tipping points is uncertain, and the Insurance Policy frame and arguments have been challenged.[22] Nonetheless, it seems prudent to extensively investigate and debate the scientific, ethical, and political arguments for using SRM as an Insurance Policy against climate emergencies.

Both the Plan B and Insurance Policy frames for SRM create, in Niebuhr's terms, temptations for sin. The seemingly ethical and reasonable motive of *buying back lost time* could create the temptation to *delay to protect wealth and power*. The worthy motive of *preparing for a climate emergency* could create the temptation to "overreach" by setting the stage for attempts at *long-term climate control*. I will first look at the temptation to delay action to protect wealth and then briefly mention the conceit of climate control.

THE TEMPTATION TO DELAY

The uses of fossil fuels are so pervasive and intertwined in the global economy that it is hard to fathom the extent and magnitude of the vested interests in keeping this form of energy at the center of the world's economies for as long as possible. Fossil fuels supply well over three-quarters of the world's commercial energy production. To state the obvious, modern civilization would collapse if fossil fuel production were to suddenly stop. Many nations—for example, some members of the Organization of Petroleum Exporting Counties (OPEC)—have economies that are completely dependent on their fossil fuel industries. Hundreds of millions of people worldwide are directly or indirectly reliant on the various fossil fuel industries. In the near term, the economic security of vast numbers of people, numerous global corporations, and many nations rest on the continued successes of the fossil fuel industry.

Temptations arise out of anxiety and insecurity when one's "position" (life, health, security, loves, wealth, power, reputation, identity, etc.) in the world is put at risk. This fact must be fully appreciated: millions of people, whose position in the world is in some way dependent on the fossil fuel industries, must feel great anxiety as discussions of moving to a low or no carbon economy become louder, more urgent, and more moralistic. To be fair, many of these same people are no doubt highly conflicted, as they also feel anxious about the serious, long-term threats of dangerous climate change. That is, they have deep concerns about the fate of the planet and the security of the lives of their children and grandchildren. Fear of losing or diminishing one's position in the world creates anxiety, and it is only natural to take actions to seek security. Further, the moralistic tone of much climate change discourse paints those in the fossil fuel industry as complicit in an evil being done to the planet, the poor, and future generations. Being vilified for working in an industry that is currently providing a necessary service to support civilization must create emotional dissonance, causing additional anxiety.

It should then be no surprise that the fossil fuel industry spends millions of dollars to influence political processes to protect their wealth and increase their security. For example, it is well publicized and documented that for decades the fossil fuel industry has directly or indirectly financially supported free-market oriented groups and think tanks whose goal has been to delay political action on climate change by casting doubt on the scientific consensus represented by the United Nation's International Panel on Climate Change reports.[23] The temptation for such dishonest and undue influence on climate change politics can be understood, even if the behavior is not condoned. If those who are heavily vested in the fossil fuel industry believe these activities could protect their financial interests and position in the world, it is easy to see why they might resist the science on climate change. Contemporary science on climate engineering might create temptations of its own sort.

THE TEMPTATION TO CONTROL THE CLIMATE

While the main focus of this chapter is the first temptation just discussed, it is worth briefly discussing the second temptation, climate control. Human control of the entire climate system would be far in the future, if it is even possible, but it is important to mention it given the topic of the temptations of climate engineering. If serious scientific research and technological development of climate engineering were to begin in earnest, knowledge of the earth's climate system and how it might react to human intervention would grow. The rate of growth of scientific knowledge and the extent and direction of technological developments cannot be predicted. One of the founding figures in philosophy of technology, Hans Jonas, writes:

> The Promethean enterprise of modern technology speaks a different language. The word "enterprise" gives the clue, and its unendingness another. We have mentioned that the effect of its innovations is disequilibrating rather than equilibrating with respect to the balance of wants and supply, always breeding its own new wants. This in itself compels the constant attention of the best minds, engaging the full capital of human ingenuity for meeting challenge after challenge and seizing the new chances.[24]

Jonas provided several important insights on the relationship between modern science, technology, human needs, and human desires. Modern technologies do not simply function as means to satisfy ends. As scientific knowledge grows and new technologies develop, those new technologies create new ends and desires. For example, when the scientists at Intel introduced the first microprocessor chip in 1971, no one could have predicted that personal computers, tablets, and the smart phones would become a necessity for so many people in 2015. This innovation and the new devices it spawned created new wants, needs, and desires. Similarly, when Watson and Crick first described the double helix structure of the DNA molecule in 1953, no one could have predicted the new ends and desires that are being created by the nascent genomic revolution, which is on the cusp of transforming medicine and agriculture. Likewise, it seems improbable that the science and technology of climate engineering will simply stop once the means of climate engineering technology satisfies the possible ends of a Plan B or an Insurance Policy. The science and technology of climate engineering are likely to follow the circular and self-reinforcing pattern described by Jonas, where new knowledge creates new technologies, which create new desires that drive the growth of knowledge and technological development in unexpected directions.

The climate engineering expert David Keith, who has consistently advocated researching SRM, concludes his essay "Engineering the Planet" with the following prediction and advice:

> Nevertheless, I judge it likely that this century will see serious debate about—and perhaps implementation of—deliberate planetary-scale engineering. Active planetary management may be an inevitable step in the evolution of a technological society, but I urge caution. We would be wise to practice walking before we try to run, to learn to minimize impacts before we try our hand at planetary engineering.[25]

The unpredictability of weather and climate are perennial sources of human anxiety and insecurity. If it were believed that the means existed to control the climate, it could create a temptation to make life more secure by doing so. Once again, as scientific knowledge grows and technological power develops, there might be a temptation to consider long-term climate control, and this temptation could lead to the sin of overreach, and overreach predictably leads to tragic results.

TEMPTATION, SIN, AND OVERARCH

Many of the things Niebuhr said about temptation and sin can be directly applied to the use of large-scale technological power such as SRM. Niebuhr writes:

> Man is insecure and involved in natural contingency; he seeks to overcome his insecurity by a will-to-power which overreaches the limits of creatureliness. Man is ignorant and involved in the limitations of a finite mind; but he pretends that he is not limited. He assumes he can gradually transcend finite limitations.[26]

The occasion for temptation to overreach lies in the fact that we are limited by our position in nature, but we are to some extent free to transcend those limitations, most significantly through scientific technology. The source of the temptation to overreach is created by the combination of the fragility of human existence, the plasticity of human limits, and the ever-growing power of scientific technology to extend those limits.

The contemporary philosopher of technology Andrew Feenberg uses hubris to describe this danger. Sounding very much like Niebuhr, he notes that the Greek word for "overweening striving" at godlike power is *hubris*.[27] Feenberg observes that hubris arises out of the finite and fallible nature of human existence. In Feenberg's secular, philosophical analysis, like in Niebuhr's theological analysis, the problem is not that we use scientific-technology to make our lives more secure, but that we extend that power beyond ethical and epistemic limits—we overreach. In Niebuhr's analysis, it is human insecurity coupled with pride and self-importance that keeps us from identifying those limits. For Feenberg, it is a failure of critical self-knowledge: we fail to follow Socrates's dictum, adopted from the god Apollo, to know thyself. Feenberg develops these ideas by asserting that we can never be like gods and know or control the universe from outside; human life and human cultures are situated within natural systems. He writes:

> Humans are indeed unique among living things in that they have the power to modify their niche to an extraordinary extent. This supports the illusion that

> they are in fact independent of the world to which they belong. But that this is an illusion is clear from the long-term consequences of ignoring all limits.[28]

In a similar vein, Crouter writes: "For Niebuhr our knowledge of the world, like our knowledge of ourselves, remains limited, even when expressed with confidence. Simply put, the complexity of things exceeds our grasp in ways that our fragile selfhood seeks to minimize and deny."[29] Humans are inescapably embedded in the subtle complex web of connections that comprise the systems in which we evolved. Because of this, when we act on those systems to make our lives more prosperous and secure, we are inevitably acting on ourselves, often in imperceptible and surprising ways. Stated differently, whatever technological power we exert on a system, to lesser or greater degrees, will in some way be exerted back on us. Modern technologies often hide this inescapable fact about our actions. Feenberg calls ignorance of such connectedness the illusion of technology.

Western civilization is to an important degree defined by scientific and technological power aimed at controlling nature to make our lives more secure. During the last few centuries there have been many amazing achievements in medicine, agriculture, industry, and elsewhere that have made the lives of hundreds of millions of people more secure from disease, famine, and want. Despite these achievements, we remain ever anxious because we are fragile, finite, and fallible creatures. This creates an ever-present temptation to overreach, as we use the increasing power of scientific technology to reduce anxiety and seek security. We slide into the sin of hubris when we overestimate our power and come to believe that we are in control through the exercise of the ever-increasing power of scientific technology. In Niebuhr's analysis of the human condition, this inescapable anxiety creates a temptation that can lead some to sin. The various types of vested interests mentioned at the start of this chapter (the 4Fs—fortune, fear, fame, and fanaticism) could each be interpreted as specific forms of anxiety. As such, the threat of the sin of overreach creates a constant risk for climate engineering governance and possible deployment.

It is important to keep in mind that for Niebuhr neither anxiety nor the use of technological power to seek security is sinful. Our ever-anxious state and the constant growth of technological power simply create the opportunity for sin. Summarizing Niebuhr's key insights on the sin of overreach or hubris, Robin Lovin writes: "We tend to see ourselves as more powerful and more righteous than we really are, and unless we correct these errors, we will be led into adventures that exceed our power and further corrupt our judgment."[30] Whether we are speaking of military power or technological power, Niebuhr's analysis of temptation and sin warns us to be highly self-critical of any assumptions we make about the extent of our power and the purity of our motives. Humans often overreach on both counts.

SRM AND THE TEMPTATION TO OVERREACH

To summarize the argument thus far, the Plan B and Insurance Policy frames and arguments for SRM research create the temptation for the sin of overreach, or hubris. Research conducted for the seemingly good motives of buying back lost time for mitigation and adaptation policies to work could create the temptation to use SRM to protect fortunes and power tied to the fossil fuel economy. Research conducted to prepare for climate emergencies could create the temptation to set the stage for the hubris of long-term climate control. It would be wise to understand the nature of these temptations and prepare for them in order to avoid slowly sliding into overreach or hubris, which could lead to a catastrophe.

SRM is a technological fix, an attempt to reframe an intractable social and political problem as an engineering puzzle. The intractable problem is, of course, the decades of ineffective cultural and political responses to climate change. SRM is an engineering puzzle in which scientists explore various ways of manipulating the earth's solar budget to keep the climate from surpassing a dangerous threshold. The appeal of technological fixes is that they take complex and seemingly insolvable social and political problems and reframe them in much simpler technical terms. By design, technological fixes naturally focus attention on the science and engineering challenges and away from political challenges. However, political factors cannot ultimately be escaped, and drawing attention away from them presents hazards.

Political decisions will ultimately determine the most important questions about SRM governance and possible deployment. For example, at the end of a report listing the scientific and technical issues involved with a climate engineering response to a climate emergency, Blackstock et al. note the central political issue: Who decides? Who decides "what climate engineering research should be pursued"? Who decides "whether or which climate engineering alternative should be deployed"? And who decides "what the 'optimal' target climate should be, both globally and regionally?"[31] These decisions will be made within the political sphere. It is here where Niebuhr's hopeful realism can be instructive. Niebuhr's analysis is clear-eyed enough to warn of the temptations to sin that climate engineering creates. But it is not so cynical as to see only vested interest, anxiety, and the will to power. His hopeful realism allows for the possibility, through sober assessments of the extent of our power and the purity of our motives, to cautiously move forward in good faith.

Most scientists who are interested in climate engineering seem interested in this line of research for humanitarian and environmental reasons. These scientists understand better than most the dangers of climate change, and have demonstrated no interest in pursuing climate engineering to protect the wealth of the fossil fuel industry.[32] Scientists who investigate SRM do not tend to view it as an alternative to strong greenhouse gas mitigation polices. However, it would be naïve for such scientists to think that their motives for working on SRM, or indeed other forms of climate engineering, will determine the ultimate directions of climate engineering politics. The obvious lesson of decades of climate change politics is that vested interests, when they possess the wealth to mount extraordinary efforts, will heavily weight the answers to the question: who decides? Niebuhr warns:

> Greed is a form of the will-to-power that has been a particularly flagrant sin in the modern era because modern technology has tempted contemporary man to overestimate the possibility and value of eliminating his insecurity in nature. Greed has become the besetting sin of bourgeois culture.[33]

So while scientists may be motivated to work on SRM (and other climate engineering projects) for environmental or humanitarian values, there is no guarantee that those values will determine how this technology might be ultimately applied.

Much can be learned about the likely directions and patterns of climate engineering politics from studying decades of climate change politics. The past behaviors of vested interests might predict that they will use their power to influence climate engineering politics. There are hints that the same constituencies that sought to delay climate policies might view SRM as a means to delay aggressive greenhouse gas reduction policies and prolong the use of fossil fuels. Writing in the journal of the free-enterprise think-tank, the *American Enterprise Institute*, Samuel Thernstron writes: “Seen in its proper light, geoengineering is potentially the key to unlock the mitigation puzzle—a way of controlling climate risks during the many decades that it will take to transform the global energy system.”[34] The question, of course, is when does buying time slide into delaying to protect wealth and power? Rex Tillerson, the Chief Executive Officer of Exxon commented that “we’ll adapt to [climate change]. It’s an engineering problem and it has engineering solutions.”[35] The simplest engineering solution for the head of the fossil fuel corporation may be to use SRM to blunt the worst effects of climate change while the use of fossil fuels is extended as long as possible.

Modern fossil fuel energy development and infrastructure requires large, long-term investments. Investors will be less willing to make such commitments if they believe that societies are transitioning away from fossil fuels. However, Holly Buck suggests that if governments fund climate engineering research and this research is perceived by politicians and negotiators as promising, while greenhouse gas emission cuts continue to stall, then investors might be more likely to take financial risks on expensive fossil fuel infrastructure.[36] Buck concludes, “The idea of geoengineering, or perceptions about its potential, could be more dangerous than the reality of it.”[37] As was noted, over the last several decades the fossil fuel industry and its various allies have become extremely adept at manipulating perceptions of climate science. There is no reason to think that the cynical "denial machine" that was aimed at undermining climate science could not be retooled to overstate the certainty and reliability of climate engineering science.

Over the last several decades of climate change politics, a large and diverse group of free-market conservatives and pro-business organizations joined forces with the fossil fuel industry to oppose strong climate change policies. They viewed these policies (such as cap and trade or carbon tax proposals) as fundamental threats to the existing order, to their way of life.[38] A complex of political groups, business groups, think tanks, and industries mounted a successful effort to delay climate change policies by attacking climate science—supposing that if people believed climate change was not a serious and imminent threat, there would be no need to act aggressively, this issue. However, as direct evidence of the threat of dangerous climate change becomes more tangible, this political strategy may stop being as effective. This situation could create the temptation to use SRM as a political tool to once again delay climate change polices to preserve the exiting order as long as possible. If the objective is to delay the move away from fossil fuels, and denial campaigns are no longer effective, then pushing discussions of SRM projects into the political limelight might be the best option.

CONFRONTING THE TEMPTATION TO OVERREACH IN SRM POLITICS

The beginning of this chapter noted that Jane C. S. Long and I identified four types of vested interest that could pose a threat to trustworthy climate engineering research and governance, the 4 Fs—fortune, fear, fame, and fanaticism. This chapter has only investigated aspects of two of the four Fs—fortune and fear. Much more work needs to be done to investigate the roles of various types of vested interest in climate engineering politics. The purpose of such investigations, and of this chapter, is not to discourage research, but to warn that it could create temptations for the sin of overreach or hubris. Given the possible scenarios described in this chapter involving fortune and fear, what kinds of actions might be taken to keep the temptations of SRM from sliding into the sin of overreach? That is a difficult question, and I do not pretend to answer it at the end of this chapter. A good place to start would be to study the strategies and behaviors of various vested interests in climate change politics. One particularly important area of study would be the uses of financial and political power to distort perceptions of the science of climate engineering.

Niebuhr's insights suggest that it will be a daunting task to keep climate engineering science from being manipulated as an instrument to gain a political advantage. Fortunately, much has been learned about ethics, power politics, and climate science communication over the last twenty years. For example, in 2005 the Yale School of Forestry and Environmental Studies held an influential conference on climate change, summarized in *Americans and Climate Change: Closing the Gap between Science and Action.*[39] The purpose of the conference was to investigate why political action to address dangerous climate change was falling so far behind our scientific understanding of the problem, and to recommend ways of closing this gap. Out of that conference grew the Yale Project on Climate Change Communication. This project has contributed to an increase in understanding public knowledge on climate change and has led to the development of effective strategies for communicating climate science. Further, there have been numerous efforts aimed at journalists to improve reporting on climate change. There has been much social science and ethics research on climate science communication and politics that could be useful in combating distortions of climate engineering science. These efforts grew out of a reaction to the manipulation of climate science and the political consequences. It would be wise to anticipate the possible distortion of climate engineering science and proactively pursue similar projects and efforts to the ones listed above targeted on climate engineering.

Finally, on the legal front, there has been a recent development of great interest.[40] The New York Attorney General, Eric Schneiderman, is investigating Exxon Mobile for deceiving its investors on the risks of climate legislation to the company's profits. More importantly, the Attorney General is investigating the fossil fuel company for deceiving the public by pursuing the agenda of climate denial. This new legal front could expand to other companies and could be joined by other states' attorneys general. If successful, these legal actions could discourage the blatant manipulation of climate engineering research to protect vested interests in the future. The fact that vested interests would not be allowed to manipulate science without the threat of legal action could help restore trust in public discourse and political processes.

There is a real danger that climate engineering science will repeat the unfortunate fate of climate science more generally and become a political tool of moral crusaders and cynical realists. However, the best way to moderate a messy and dangerous political debate over climate engineering would be to effect public policies that reduce greenhouse gas emissions. Earlier it was noted that Niebuhr's Christian realism is optimistic enough to encourage sustained efforts at building cooperation but realistic enough to know that cooperation will always be imperfect and tainted by self-interest. Climate change presents an extraordinarily difficult challenge that must be approached with modesty and humility, but also with courage and resolve. One must equally avoid the naïve idealism of overwrought moralistic crusades to save the planet and the bitter realism of believing that nations are only capable of acting in their narrow self-interest.

Nowhere is the need to avoid the extremes of moralistic crusades and cynical realism, to embrace a hopeful realism that builds cooperation, more present than climate change politics. If we cannot do this, then some form of risky climate engineering might be the best alternative in an increasingly limited and bad set of options.

NOTES

1. Jane C. S. Long and Dane Scott, "Vested Interest and Geoengineering Research," *Issues in Science and Technology* 29 (Spring 2013): 45–52.
2. Ibid., 45.
3. Richard Crouter, *Reinhold Niebuhr, On Politics, Religion and Christian Faith* (Oxford: Oxford University Press, 2010), 42.
4. Ben Hale, "The World That Would Have Been: Moral Hazard Arguments against Geoengineering," in *Engineering the Climate*, ed. Christopher Preston (Lanhan, MD: Lexington Books, 2012), 113–32.
5. Crouter, *Reinhold Niebuhr, On Politics, Religion and Christian Faith*, 41.
6. Reinhold Niebuhr, *The Irony of American History* (Chicago: The University of Chicago Press, 2008), 134.
7. David Brooks, "Obama, Gospel and Verse," *New York Times*, April 26, 2007.
8. Anatol Lieven, "Realism and Progress, Niebuhr's Thought and Contemporary Challenges," in *Reinhold Niebuhr and Contemporary Politics: God and Power*, ed. Richard Harries and Stephen Platten (Oxford: Oxford University Press, 2010), 169.
9. Reinhold Niebuhr, *Christian Realism and Political Problems* (New York: Scribner's, 1953), 137.
10. Lieven, "Realism and Progress, Niebuhr's Thought and Contemporary Challenges," 169.
11. Ibid., 170.
12. Jean Bethke Elshtain, "Niebuhr's 'Nature of Man' and Christian Realism," in *Reinhold Niebuhr and Contemporary Politics: God and Power*, ed. Richard Harries and Stephen Platten (Oxford: Oxford University Press, 2010), 51–52.
13. David Keith, Edward Parsons, and M. Granger Morgan, "Research on Global Sun Block Needed Now," *Nature* 462: 426–27.
14. Katharine Ricke, M. Granger Morgan, Jay Apt, David Victor, and John Steinbruner, "Unilateral Geoengineering, Non-technical Briefing Notes, Workshop at the Council on Foreign Relations, Washington DC," May 5 (2008), accessed December 3, 2015, http://www.cfr.org/projects/world/geoengineering-workshop-on-unilateral-planetary-scale-geoengineering/pr1364.
15. Alan Robock, Luke Oman, and Georgiy L. Stenchikov, "Regional Climate Responses to Geoengineering with Tropical and Arctic SO2 Injections," *Journal of Geophysical Research* 113 (2008): 1–15.
16. There are numerous competing frames for climate engineering. For example, in this volume Hartman offers insightful analyses of various frames. She offers the fruitful suggestion to frame geoengineering in terms of medicine and healing.
17. Kevin Anderson, "Talks in the City of Light Generate More Heat," *Nature* 528 (December 24, 2015): 437.
18. Intergovernmental Panel on Climate Change, Working Group III, accessed January 26, 2016, http://www.ipcc-wg3.de/assessment-reports.

19. Oliver Milman, "James Hansen, Father of Climate Change Awareness, Calls Paris Talks 'a Fraud,'" *The Guardian,* December 12, 2015.

20. Stephan H. Schneider, "Geoengineering: Could—or Should—We Do It?" *Climatic Change* 33 (1996): 291–302.

21. Timothy Lenton, "Early Warning of Climate Tipping Points," *Nature Climate Change* 1 (2011): 201–9.

22. Joshua B. Horton, "The Emergency Framing of Solar Geoengineering: Time for a Different Approach," *The Anthropocene Review* 2 (2015): 147–51; Jana Sillmann, Timothy M. Lenton, Anders Levermann, Konrad Ott, Mike Hulme, François Benduhn, and Joshua B. Horton, "Climate Emergencies Do Not Justify Engineering the Climate," *Nature Climate Change* 5 (2015): 290–92.

23. Naomi Oreskes and Eric M. Conway, *Merchants of Doubt* (New York: Bloomsbury Press, 2011).

24. Hans Jonas, "Toward a Philosophy of Technology," *The Hastings Center Report* 9 (1979): 38.

25. David Keith, "Engineering the Planet," in *Climate Change Science and Policy*, eds. Stephen H. Schneider, Armin Rosencranz, Michael Mastrandrea, and Kirsten Kuntz-Duriseti (Washington, DC: Island Press, 2010), 501.

26. Reinhold Niebuhr, "The Pride of Power," in *Readings in Christian Theology*, ed. Peter Hodgson, Robert Crafts, and Robert H. King (Minneapolis, MN: Fortress Press, 1985), 178.

27. Andrew Feenberg, "Technology and Human Finite," (Unpublished, 2015).

28. Ibid.

29. Crouter, *Reinhold Niebuhr, On Politics, Religion and Christian Faith*, 130.

30. Robin Lovin, "Christian Realism for the Twenty-first Century," *Journal of Religious Ethics* 37, 4 (2009): 669.

31. J. J. Blackstock, D. S. Battisti, K. Caldeira, D. M. Eardley, J. I. Katz, D. W. Keith, A. A. N. Patrinos, D. P. Schrag, R. H. Socolow, and S. E. Koonin, "Climate Engineering Responses to Climate Emergencies," November (2009), accessed December 3, 2015, http://arxiv.org/pdf/0907.5140.

32. Holly Jean Buck, "Geoengineering: Re-making Climate for Profit or Humanitarian Intervention?" *Development and Change* 43, 1 (2012): 259.

33. Niebuhr, "The Pride and the Power," 178.

34. Samuel Thernstron, "What Role for Geoengineering," *The American: The Journal of the American Enterprise Institute*, March 2010, accessed June 20, 2014, http://www.aei.org/publication/what-role-for-geoengineering/.

35. Matt Dailey, "Exxon CEO Calls Climate Change Engineering Problem," *Reuters*, 2007, accessed June 27, 2012, http://www.reuters.com/article/2012/06/27/us-exxon-climate-idUSBRE85Q1C820120627.

36. Buck, "Geoengineering: Re-making Climate for Profit or Humanitarian Intervention?"

37. Ibid., 258.

38. Riley McCright and Aaron Dunlap, "The Politicization of Climate Change and the Polarization of American Public's Views of Global Warming," *The Sociological Quarterly* 52 (2011), 155–194; Naomi Klein, "Capitalism vs. the Climate," *The Nation*, November 9, 2011.

39. Daniel R. Abbasi, *Americans and Climate Change: Closing the Gap between Science and Action* (New Haven, CT: Yale School of Forestry and Environmental Studies, 2006).

40. Justin Gillis and Clifford Kraus, "Exxon Mobile Investigated for Possible Climate Change Lies by New York Attorney General," *New York Times,* November 5, 2015.

BIBLIOGRAPHY

Abbasi, Daniel R. *Americans and Climate Change: Closing the Gap between Science and Action.* New Haven, CT: Yale School of Forestry and Environmental Studies, 2006.

Anderson, Kevin. "Talks in the City of Light Generate More Heat." *Nature* 528 (December 24, 2015): 437.

Blackstock, J. J., D. S. Battisti, K. Caldeira, D. M. Eardley, J. I. Katz, D. W. Keith, A. A. N. Patrinos, D. P. Schrag, R. H. Socolow, and S. E. Koonin. "Climate Engineering Responses to Climate Emergencies." November (2009). Accessed December 3, 2015. http://arxiv.org/pdf/0907.5140.

Brooks, David. "Obama, Gospel and Verse." *New York Times*. April 26 (2007). Accessed January 26, 2016. http://www.nytimes.com/2007/04/26/opinion/26brooks.html?_r=0.

Buck, Holly Jean. "Geoengineering: Re-making Climate for Profit or Humanitarian Intervention?" *Development and Change* 43, 1 (2012): 253–70.

Crouter, Richard. *Reinhold Niebuhr, On Politics, Religion and Christian Faith.* Oxford: Oxford University Press, 2010.

Dailey, Matt. "Exxon CEO Calls Climate Change Engineering Problem." *Reuters*. (2007) Accessed June 27, 2012. http://www.reuters.com/article/2012/06/27/us-exxon-climate-idUSBRE85Q1C820120627.

Elshtain, Jean Bethke. "Niebuhr's 'Nature of Man' and Christian Realism." In *Reinhold Niebuhr and Contemporary Politics: God and Power*, edited by Richard Harris and Stephen Platten (Oxford: Oxford University Press, 2010), 42–57.

Feenberg, Andrew. "Technology and Human Finite." Unpublished, 2015.

Gillis, Justin, and Clifford Kraus. "Exxon Mobile Investigated for Possible Climate Change Lies by New York Attorney General." *New York Times*. November 5 (2015). Accessed January 26, 2016. http://www.nytimes.com/2015/11/06/science/exxon-mobil-under-investigation-in-new-york-over-climate-statements.html.

Hale, Ben. "The World That Would Have Been: Moral Hazard Arguments against Geoengineering." In *Engineering the Climate*, edited by Christopher Preston (Lanham, MD: Lexington Books, 2012), 113–32.

Horton, Joshua B. "The Emergency Framing of Solar Geoengineering: Time for a Different Approach." *The Anthropocene Review* 2 (2015): 147–51.

Jonas, Hans. "Toward a Philosophy of Technology." *The Hastings Center Report* 9 (1979): 34–43.

Keith, David. "Engineering the Planet." In *Climate Change Science and Policy*, edited by Stephen H. Schneider, Armin Rosencranz, Michael Mastrandrea, and Kirsten Kuntz-Duriseti (Washington, DC: Island Press, 2010), 494–501.

Keith, David, Edward Parsons, and M. Granger Morgan. "Research on Global Sun Block Needed Now." *Nature* 462: 426–27.

Klein, Naomi. "Capitalism vs. the Climate." *The Nation*. November 9 (2011). Accessed January 26, 2016. http://www.thenation.com/article/capitalism-vs-climate/.

Lenton, Timothy. "Early Warning of Climate Tipping Points." *Nature Climate Change* 1 (2011): 201–9.

Lieven, Anatol. "Realism and Progress, Niebuhr's Thought and Contemporary Challenges." In *Reinhold Niebuhr and Contemporary Politics: God and Power*, edited by Richard Harris and Stephen Platten (Oxford: Oxford University Press, 2010), 169–81.

Long, Jane C., and Dane Scott. "Vested Interest and Geoengineering Research." *Issues in Science and Technology* 29 (Spring 2013): 45–52.

Lovin, Andrew. "Christian Realism for the 21st Century." *The Journal of Religious Ethics* 37, 4 (2000): 669–82.

McCright, Riley, and Aaron Dunlap. "The Politicization of Climate Change and the Polarization of American Public's Views of Global Warming." *The Sociological Quarterly* 52 (2011): 155–94.

Milman, Oliver. "James Hansen, Father of Climate Change Awareness, Calls Paris Talks 'a Fraud'." *The Guardian*. December 12 (2015). Accessed January 26, 2016. http://www.theguardian.com/environment/2015/dec/12/james-hansen-climate-change-paris-talks-fraud.

Niebuhr, Reinhold. *Christian Realism and Political Problems*. New York: Scribner's, 1953.

Niebuhr, Reinhold. *The Irony of American History.* Chicago: The University of Chicago Press, 2008.

Niebuhr, Reinhold. "The Pride of Power." In *Readings in Christian Theology*, edited by Peter Hodgson, Robert Crafts, and Robert H. King (Minneapolis, MN: Fortress Press, 1985), 112–17.

Oreskes, Naomi, and Eric M. Conway. *Merchants of Doubt.* New York: Bloomsbury Press, 2011.

Ricke, Katharine, M. Granger Morgan, Jay Apt, David Victor, and John Steinbruner. "Unilateral Geoengineering, Non-technical Briefing Notes, Workshop at the Council on Foreign Relations, Washington DC." May 5 (2008). Accessed December 3, 2015. http://www.cfr.org/projects/world/geoengineering-workshop-on-unilateral-planetary-scale-geoengineering/pr1364.

Robock, Alan, Luke Oman, and Georgiy L. Stenchikov. "Regional Climate Responses to Geoengineering with Tropical and Arctic SO2 Injections." *Journal of Geophysical Research* 113 (2008): 1–15.

Schneider, Stephan H. "Geoengineering: Could—or Should—We Do It?" *Climatic Change* 33 (1996): 291–302.

Sillmann, Jana, Timothy M. Lenton, Anders Levermann, Konrad Ott, Mike Hulme, François Benduhn, and Joshua B. Horton. "Climate Emergencies Do Not Justify Engineering the Climate." *Nature Climate Change* 5 (2015): 290–92.

Thernstron, Samuel. "What Role for Geoengineering." *The American: The Journal of the American Enterprise Institute.* March (2010). Accessed June 20, 2014. http://www.aei.org/publication/what-role-for-geoengineering/.

Chapter Four

Real Presence: Process Theological Perspectives on Geoengineering the Body of God

Marit Trelstad

At first glance, geoengineering may appear to be a topic beyond the average Christian. Its name suggests a discussion for scientists and engineers. What has Christianity to say about intentional, large-scale climate intervention? The Bible describes humans as responsible for the care of all the creatures and plants of the earth, as early as the first chapter of Genesis. But there are no direct mandates or guidelines in scripture about geoengineering itself, and Christian historical teachings and theology do not clarify the means by which we should tend to creation. Likewise, geoengineering may not appear to need Christian theological reflection. We do not need to invoke Christian-sounding ideas or metaphors in order to justify a particular position on an issue if this does not offer any novel insight. Nor do we need to put Christian language on an existent, perfectly sensible secular debate.

Theology, at its best, is able to exhume and examine our foundational beliefs for their coherence and sources. It offers a systematic examination of ideas implicit in a tradition or culture. Theology also seeks to use these foundations as constructive guideposts for understanding and engaging a variety of social and religious matters. Thus, Christian theology offers an examination of the values and worldviews behind various approaches to geoengineering as well as offering values nonnative to other fields.

This chapter examines how the Christian theological tradition and process theology offer tools for reflection to guide decisions over large-scale interventions for the care of earth and neighbor. Christian theology offers two interrelated beliefs to the conversation about geoengineering: the earth as the body of God and the communal nature of this body. Process theology and philosophy also assert that the earth is God's body and that all of creation is deeply and internally interconnected. These two key beliefs of Christian theology and philosophy—affirming the sacredness of the world and the deep interconnectivity of life—bring new perspectives to discussions of geo-engineering. The earth is understood by Christian theology and process philosophy as a living organism rather than a mechanism for our own wealth and ends. In care for the body of God, or the body of Christ as the Church, Christians are also called into involvement in issues of international, large-scale geoengineering which impact human and nonhuman bodies. Christians do this with the "greatest commandments" in mind, to love God and to love our neighbor as ourselves, in this time of eco-crisis. The beliefs and principles from the Christian tradition add a different set of values to the conversation. This is necessary so that global decisions are not determined solely by big businesses or powerful nations who may only have economic profit or market growth as their key guiding principles.

THE BODY OF GOD

Much of the Christian tradition points to the nature of all bodies, including the planet as a whole, as members of the body of God. The body of Christ and God's body are not simply metaphors but tangible realities, according to past and present Christian theology and practice. While other authors in this volume work with metaphorical language and its impact, this chapter looks to affirm God's real presence in creation.

Creation as God's body, and the church as Christ's body, are different yet connected concepts in the theology. Both the body of God and the body of Christ point to the interrelated and tangible form of God known in the church and, ultimately, all of creation and therefore both will be referenced throughout. The Body of Christ has traditionally represented the church, comprised of its members. This claim can therefore speak to the ethical necessity of caring for our human neighbors' needs in the context of climate change. The Body of God, for the purposes of the chapter, will be used to talk about God's real presence in all of creation. The earth is conceived as God's real body: sacramentally, philosophically from a realist perspective, and philosophically from a rationalist perspective. These are spelled out in detail since most Christians may not have considered that the earth can literally embody God's real presence. In fact, much of popular Christianity suggests that God is only fully present in an afterlife that is separate from this world.

This chapter is certainly not the first theological work that has supported the concept of the earth as the body of God. The most well known comes from the 1993 work of theologian Sallie McFague who published a book entitled *The Body of God: An Ecological Theology*. Arguing that metaphors shape ethical action, she proposes that the Earth can be understood to be God's Body. In *The Body of God*, she proposes a panentheistic theology where God is not equated to the natural world but also not seen as separate and "above" it.[1] She writes that Christianity's doctrines convey a deep appreciation of bodies through theological and biblical emphases on God as creator, the incarnation, and doctrines on the Holy Spirit. The Christian tradition embodies this commitment to embodiment through theologies of creation, sacraments, sabbath, ministries of feeding and healing, and more. Indeed, there are so many affirmations of God's deep relation to physical reality that McFague proposes the world could be understood to be held within the divine nature; in this way, the world is the body of God. She writes:

> We might realize what this tradition has told us, although often shied away from embracing unreservedly: we live and move and have our being *in* God. We might see ourselves and everything else as the living body of God. . . . If God is available to us in our bodies, then bodies become special.[2]

Additionally, she writes, "The world (universe) as God's body is also, then, a radicalization of divine immanence, for God is not present to us in just one place (Jesus of Nazareth, although also and especially, paradigmatically there), but in and through all bodies, the bodies of the sun and moon, trees and rivers, animals, and people."[3]

Throughout the book, however, McFague repeatedly claims that the earth is not *really* God's body but that this metaphor helps us understand theological truths of God's transcendence, immanence and the role of bodies in Christian faith. "In other words, we are dealing here with a model or metaphor, not a description: the universe as God's body is a rich, suggestive way to radicalize the glory, awesomeness, the beyond-all-imagining power and mystery of God in a way that at the time radicalizes the nearness, the availability, the physicality of divine immanence."[4]

Building on the insights of McFague's work, this chapter argues that Christian theology also offers a more literal sense of the earth as God's body, and this may form a stronger basis for an ecological ethic capable of informing discussions over geoengineering. For example, we may imagine the difference real versus symbolic bodies make when it comes to Jesus's command to feed the hungry, heal the sick, give to those who are impoverished, or visit the imprisoned. Jesus claims that we do these things to him when we do them for a fellow human. "Truly I tell you, just as you did it to one of the least of these who are members of my family, you did it to me" (Matthew 25:40). Does he mean that these people are symbolically him? Really him? Or some other meaning between these two? Does it matter whether my neighbor is Christ to me or if my neighbor is Christlike? Likewise, when Christians say God is Love, do they mean this metaphorically? This chapter begs us to consider the earth as God's body in a tangible, real sense and offers that the resources to do so are abundant. It also offers that a firmer theological connection to reality holds us more accountable to the earth and our neighbor's bodies because they are the knowable, present body of God.

SACRAMENTS AND REAL PRESENCE

Both the title of this chapter and this section allude to sacramental language and theological debates in the Christian tradition. When Jesus is serving his disciples the Last Supper, he offers them the bread and says "This is my body"[5] (Lk 22:19, Mt 26:26, Mk 14:22) and Paul's letter to the church at Corinth reports that Jesus said, "This is my body that is for you. Do this in remembrance of me" (1 Cor. 11:24). After this, Jesus offered them the cup of wine with the words, "This is my blood" (Mark 14:24, Matthew 26:28). Church tradition keeps this alive in the practice of communion where, in eating and drinking the sacraments of bread and wine, one becomes literally incorporated into the body of Christ. From the earliest days of Christianity, the church was referred to as the living body of Christ.

In light of contemporary ecological crises, some theologians have argued that it is not only the church that is the physical embodiment of Christ but, more expansively, all of creation. Larry Rasmussen, for example, argues that Christians need to think of the world in sacramental terms, extending the body of Christ to the nonhuman and marginalized natural world.[6] The bodies of all people and all of creation are the work of a loving creator who wishes for the thriving of all the earth. But is God, as Christ, really present in the physical bodies of church members and all of creation or is this a symbol or metaphor used to understand God's presence today? Can we claim the earth is literally God's body?

The Reformation argument over God's real or symbolic presence in the sacraments is reminiscent of this contemporary question. The sixteenth century Catholic Church held that the true substance of the bread and wine changed literally into the body and blood of Christ, its essential nature or substance is changed by the rite of the Eucharist. While the taste, feel, and look of the bread and wine (its "accidental nature") did not change, its essential nature did, and thus the bread and wine really were Christ's body and blood. An analogy of essential and accidental natures can be found today. For example, it is often said that a person's eye, skin, and hair color are external characteristics (accidental nature) that do not define "who they really are" (essential nature). The Catholic Church claimed that the bread and wine experienced a "transubstantiation" (changed substance) whereby the ordinary elements were essentially transformed into the body and blood of Christ.[7]

Early Protestant leaders debated this matter heatedly, and it became the divisive issue between Martin Luther and Ulrich Zwingli, the original founder of the Reformed tradition. At the Colloquy at Marburg, Germany, in 1529, Protestant theologians debated the nature of Christ's presence in the sacraments. Neither side of the early Protestant debate agreed with the Roman Catholic belief in transubstantiation, though Luther clung to a sense of Christ's real presence in the sacraments while Zwingli held that the bread and wine were symbolically Christ's body and blood. Luther's own understanding of the sacraments was that Christ was truly present in the bread and wine, and the subsequent Lutheran tradition holds that Christ is "in with and under" altogether infused and intermingled with the physical elements of the sacraments. Luther also held that Christ was, after the resurrection, both at the right hand of God and yet also present to all of creation. Carl Braaten explains,

> In this way, says Luther, all the universe is but one place for God, to which God is present, and is one place also for the risen Christ, whose location is precisely God's right hand of omnipresence. Or a person may be somewhere in that he or she is available there, intendable and addressable there. In this way, says Luther, Christ's body is where the bread and cup are, and this place can be any place, in that all places are one to Christ.[8]

Later Lutheran tradition argued that Christ is truly present wherever the gospel is present because the Word of God conveyed in preaching, scripture, and sacraments is the embodiment of the resurrected Christ. So, just as the Spirit of God present in the world is not symbolic but a real presence, thus God is present in the sacraments and beyond. In 1535, a middle-aged Luther expresses this sentiment in his "Lectures on Genesis": "God speaks true and existent realities, thus sun, moon, earth . . . you [and] I . . . are all words of God."[9] Other Christian theologians discuss God's real presence in all aspects of the Trinity: as Creator, Christ, and Spirit. As Creator, God is the ground of existence. As Christ, God becomes present and embodied through the historical person of Jesus, who is the incarnate Word of God present in the neighbor and stranger in need. The Holy Spirit is the advocating presence of God, living today and infusing all of creation. Some of these claims may be metaphorical or philosophical in nature, but other parts of the Christian tradition claim God's real presence, embodied in physical elements.

Contemporary theologian Rosemary Radford Ruether engages the sacramental tradition in her book *Gaia and God.* She argues that Christian commitment to earth's healing has been a part of religious thought and practice throughout the history of Christianity. Ecological commitments are carried in the theological themes of covenant, creation, and sacrament. In her section on sacraments, she researches the roots of the Jewish and Christian traditions that discuss God's presence in the physical world. Ruether explains Christianity's traditional language of a "cosmic Christology" in which the whole earth embodies the divine. She claims that Western Christianity ignored the tradition's emphasis on the wider natural world and focused primarily on the salvation of humans, particularly after the medieval and Reformation periods. Nonetheless, in both Judaism and Christianity, God is actively present in the world as creative source. To illustrate this, she discusses the work of a major early theologian, Irenaeus:

> For Irenaeus, creation is itself an incarnation of the Word and Spirit of God, as the ontological ground of bodily existence. The incarnation of the historical Christ is the renewal of this divine power underlying creation. In the incarnation divine power permeates bodily nature in a yet deeper way, so that the bodily becomes the sacramental bearer of the divine, and the divine deifies the body.[10]

Irenaeus's reference to God as the "ground of bodily existence" is perhaps the most foundational understanding of God throughout Christian theology, evident in theological works ancient and contemporary. For example, medieval theologian Thomas Aquinas claimed that God was the very act of existence and so all that exists participates in God to some degree. God was the foundation as source, sustainer, and redeemer of creation. As we can see, some strains of Christian theology deeply knit God's very being into the physical world. God's sacramental presence reveals God as omnipresent in the physical world. There is a deep tradition that claims that God is a real, embodied presence, in particular physical elements or deeply within all creation at its very foundation. Either way, the sacramental traditions and practices of the Christian tradition emphasize the embodiment and presence of God in creation.

WHITEHEAD'S PHILOSOPHICAL REALISM

Along with the Christian tradition, process philosophy and theology have affirmed that the universe is the body of God. There are two different approaches to this topic, taken by two of the early founders of this branch of thought: Alfred North Whitehead and Charles Hartshorne. Whitehead's approach was through the lens of ontology, a discussion of what comprises the basis of existence itself. Hartshorne approached the subject of the world as God's body through philosophical logic. Both affirmed that there was a philosophical ground by which one could conceive of the physical, natural world as the body or physical aspect of God.

Whitehead, the originator of process thought, developed a branch of philosophical realism, committed to basing all philosophical ideas on concrete reality. One begins philosophical inquiry with observation of reality, then one engages concepts by which to understand that reality, and then one checks those speculations against reality. He likens the philosophical process to an airplane that begins on the ground, is airborne, but then must return to the earth. Speculation is checked by its ability to best describe real processes. For example, Whitehead began with what may seem the simplest observations: that everything changes and everything is interrelated. As a physicist, he also incorporated the knowledge from quantum physics that the most fundamental level of matter is energy. From these basic premises, he began to develop a philosophy. This method is important to emphasize because it illustrates that Whitehead was not interested in language, metaphors, or concepts as the building blocks of reality. He wanted to describe, philosophically, the processes of the real world. Indeed, his extremely difficult major book is titled just that: *Process and Reality*.[11]

For the purposes of this chapter we need to understand Whitehead's concepts of God and relationality. Whitehead posited that there must be a source of the new possibilities present in each moment. Since we can see change, we know the present inherits the influence of the past, but the present also incorporates new, creative ways to be. He argues that the possibilities for novelty in each new moment must come from a source and, by the end of his philosophy, he states that this source could be called God since it has some characteristics that are also commensurate with the Christian tradition: the source of possibility is ever present, ever giving, and always in relationship with the world. In the last chapter of *Process and Reality*, Whitehead discusses how God offers lures toward the best possibilities for each moment and human and physical realities incorporate those potentialities according to their own history, evolution, and type of being.

It is probably easiest to imagine the relation of the world and God as a perpetual dialogue. God begins by imagining what is possible and offering it to the world's process of becoming. The world repeats some aspects of the past and incorporates some possibilities for change into its own concrete reality and then it offers its own influence to future moments and to God. God receives what the world has become and moves on to offer the next moment possibilities that are now relevant, based on what has just happened. The diagram in figure 4.1 is a simplified illustration of Whitehead's model where God is the source of Creativity:

Whitehead described this conception of God as having two "poles." In each moment, God begins God's experience in the *conceptual* pole, offering ideas or propositions for what it possible. These propositions include God's "lure" for what is the best possibility, but the world has freedom whether to embody this or not. God's experience of that instant "ends" in the *physical* pole, feeling what the concrete, physical reality has become and receiving it into God's own being. In this way, God perpetually receives and feels the world, responding with love and attunement to each moment. God then is

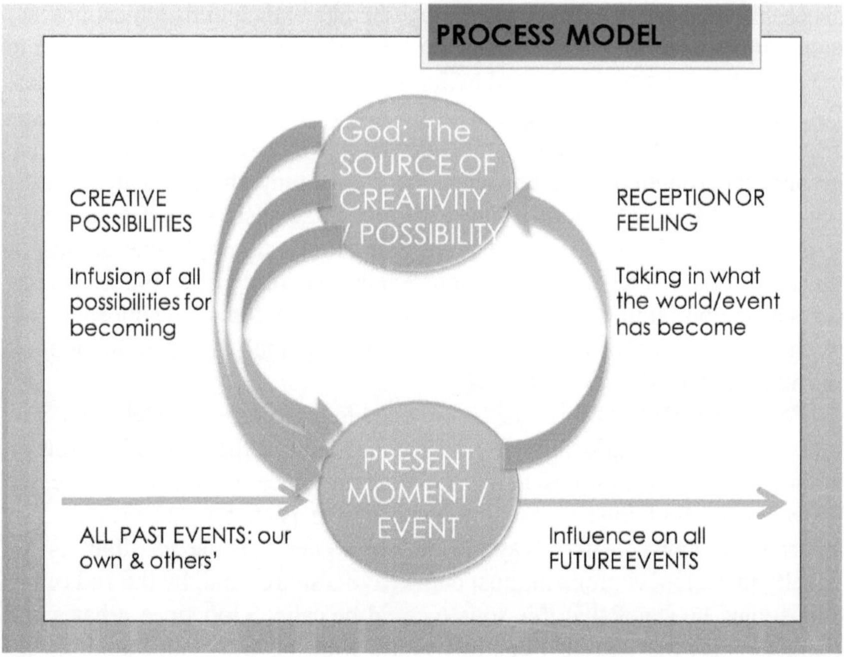

Figure 4.1.

able to offer relevant possibilities for subsequent moments because God is always related to the world and feeling it intimately.

After Whitehead's initial proposal of process philosophy, theologians took hold of this model to explain how it illustrates long-held tenets of Christian theology. For example, one of the very first ecotheologies came from theologian John B. Cobb Jr. and biologist Charles Birch who developed the idea of God as source of creativity and life.[12] As the source of Creative Life, God urges the world on to ever new forms of complexity and diverse beauty. Human actions can embody this beauty or run the other direction, serving the destruction of our ecosystems, a diminishment of life and beauty.

In Whitehead's model, there are two ways in which the world is the body of God. First, God has no body without the concrete world. The Christian claim about the person of Jesus makes sense in this schema. Jesus's incarnation of God is due to Jesus's perfect embodiment of God's best possibilities at all moments of his life—being full of the spirit of God, in biblical language. The historical person of Jesus embodied God. But, to a lesser extent, the whole world embodies God in that it becomes a concrete actuality of possibilities God has offered. While God can offer better and best possibilities, God can only *influence* but cannot *control* the choices of the world and its evolution. So sometimes we embody the very worst possibilities, based on our own experience, relationships, and history. To the extent one embodies God's loving aim and best hope for the world, one can say that one is following the way of Christ and giving physical manifestation to the vision of God for the world. The Christian tradition has called this vision the Kingdom or Reign of God. Baptismal language often calls Christians to be participants in this present and future aim of God, to be people who live out the kingdom of God and its vision of peace and justice. In these ways, the world is the concrete reality of God to the extent that it aligns itself with God's loving intent for its communal thriving.

There is a second way that Whitehead's model describes the earth as God's physical aspect. Each completed moment or event is received by God and brought into God's own reality, forever knit into the being of God. In Marjorie Suchocki's book *The End of Evil*,[13] she describes how God incorporates the world, feeling it in all its beauty and horror, and transforming it by holding it within God's self forever. Process theology here proposes a model where the world incorporates God in real matter, real moments; it is fair to say that God is the source and foundation of reality's processes. The world is the physical nature of God's conceptual possibilities. Also, the world is God's body in the sense that God takes each moment into Godself, feeling it perfectly and everlastingly even while our own physical natures ultimately perish.

HARTSHORNE'S RATIONALISM

Charles Hartshorne approaches the idea of the physical and conceptual nature of God from a different angle. He builds on the work of Whitehead and proposes a panentheist model of God. As he writes, classical theism proposes the world and God as two separate entities. Pantheism (literally translated as "all god") is the opposite, where the world is equated to God or the divine. A third option, pan*en*theism, claims that the world is within God; God holds the world within God's own being. Hartshorne proposes panentheism to be true based on simple logic. If God is to be the greatest possible being that can be conceived, God cannot be separate from the world or else there is always existence and knowledge beyond God. For example, he says that the greatest possible subset of two items, A and R, would be neither A nor R independently. Rather (A + R) would be the greatest possible subset. Likewise God's own nature (A) plus the World (R) would be the greatest possible subset. He proposes that God has an absolute, or independent, nature which is eternal and unchanging. But God also has a relative nature, which is related to and inclusive of the whole world. The absolute aspect of God is like the yolk of an egg, which is both distinct and a part of the whole.

Hartshorne proposed that these two natures of God are held within one being, one that holds both the supernatural (beyond nature) and the natural world. Thus, he alighted on the idea that the earth was the body of God even while God also has God's own independent nature. He describes God as "surrelative" since God is relative or related to all in a way that only God can be. While humans are interrelated to and composed of some relations, God is related to all things perfectly, in an unsurpassable manner. In each developing moment, then, God both changes (in God's relative aspect) and remains unchanged (in God's absolute aspect) by the world. God also contains and yet transcends each moment. Hartshorne writes:

> The total actual state of deity now, as surrelative to the present universe, has nothing outside itself, and in that sense is the All. But the individual essence of deity (what makes God God, or the divine divine) is utterly independent of this All, since any other possible all (and there are infinite possibilities of different totalities) would have been compatible with this essence. The divine personal essence in this fashion infinitely transcends the de facto totality, and every moment a partly new totality contains and embodies the essence.[14]

Quite the opposite of what he deems to be the dominant images of God in classical theology, God is related to all and feels all of existence perfectly. In Hartshorne's rationalist process philosophy, to be related is a higher form of existence than independence, so this sensitivity to all is God's greatest power. Independence requires only knowledge of oneself, but relativity requires knowledge of both oneself and the other; and to be perfectly relative to all would entail perfect knowledge of all existence. Hartshorne claims that to feel, to know, and to respond adequately to all is the meaning of God's omniscience. He explains that tyrants rule without knowledge, care, or receptivity to their subject's feelings or experience. If God is love, and love requires an openness to the response of the other, God cannot be unchanged by the world's experience. God's perfect love requires a different understanding of divine nature.

At the end of his 1948 book *The Divine Relativity: A Social Conception of God*, Hartshorne has a section entitled "Some Practical Applications." Here he ponders how world history and church tradition would have been different if Christians had understood compassion and relatedness as the highest divine powers rather than omnipotence without feeling. He writes that it could have, perhaps, avoided "power worship" which prizes God's power to influence, unilaterally, over God's sensitivity to the world. He states that this power worship has influenced theological, political, and social ideas of what it means to have "authority" in the most negative ways. He also posits that divorcing God's being from the earth has led Christians to prize "otherworldliness—the flight from the one task we surely face, that of human welfare on earth, to a questionable one, the winning of a heavenly passport."[15]

Clearly, Hartshorne claims the world to be within the nature of God though it is not equal to the totality of God. God includes both transcendence and independence. The world's fate is a divine matter in a literal sense, for both Hartshorne and Whitehead. God and the physical world, for better and for worse, are interconnected and mutually informing. God and the world are deeply interconnected and related and, for both Hartshorne and Whitehead, God feels and incorporates the world's experiences, weaving them into the very nature of God.

GEOENGINEERING AND THE BODY OF GOD

Christian theological claims about the Body of God may influence our perspectives on geoengineering. For one, understanding the real presence of God to be with, in, and under all of creation may raise our own investment in the health of the nonhuman world. In some ways, this is consistent with the argument Laura Hartman makes in chapter 7 of this volume, which develops the metaphor of the earth as a body in need of healing. But this chapter has made a more literal claim about the earth as God's body. Similar to Jesus's preaching that serving the neighbor is truly serving him, care of the earth becomes equivalent to care for the embodied divine which infuses all. It brings metaphors home, testing and grounding them by their ability to correspond to reality. God's body is not a concept or an idea but a tangible reality in need of care. In a similar vein, a slogan proposed by several Christians years ago declared: "The Body of Christ has AIDS." This called members of the church to see AIDS as an issue of our own real, corporate body and to respond with care. The grounding in reality and tangible bodies serves as a call to ethical action. Whitehead, you will recall, also insisted that abstract concepts have a correlate reality so that our abstractions remained grounded. He said that ideas are important for creativity and change, but at some point, they need to touch down to reality and be corrected or checked by them. Likewise, John Cobb has evoked this concept of Whitehead's to challenge economic theory that does not account for deterioration of nature or sustainability in its calculation of growth.

Let me give a concrete example. In my walks, I pass a lot of garbage, particularly plastic, that will be washed into the waters of Puget Sound by the next rainfall. Plastic garbage on the body of God is simply unacceptable, and so I find myself stooping to pick up bits and pieces along the way. This is not because I am a good person, but because I feel called to care for the bodies and lives entrusted to me. Working to understand the physical world's participation in God and Christian responsibility to care for the body of God engages one in imagining the systematic impact of the smallest human actions, let alone some of the large-scale proposals of geoengineering. It calls for us to do careful work regardless of our moral qualities.

Respect and religious reverence for the earth have permeated the Christian tradition from its start. When we hold such reverence, and consider that the earth may be one way in which God is embodied, our decisions may be based on deeper, more fundamental cares rather than exciting technology or financial and national benefit. It may invite us to situate our own self-interest within a greater body of the human and nonhuman world. Jesus seemed to think that his followers' actions may change if they considered the poor, weak, hungry people around them *as him*. Likewise, when our own bodies are considered as part of God's body in the world, we may treat them with greater respect. I Corinthians 6:19 states, "Or do you not know that your body is a temple of the Holy Spirit within you, which you have from God, and that you are not your own?" It is vital to take the care of bodies seriously—personally and globally, human and nonhuman—as a part of one's life with God. Such reverence of the physical world provokes us to invest in its care and to consider decisions regarding it within the context of how one would act in relation to God or Christ. It also becomes the work of all people, not a few engineers or scientists or government leaders who may or may not understand the religious value of creation which Christianity affirms from many different angles. We all have a stake in the conversation and are called to be involved. In addition, the body of Christ as the church has preferences that need to be considered. The needs of disadvantaged and suffering members are the ones that are primary. And the welfare of the enemy and one's loved ones need both be factored. No doubt this is a tough set of expectations to guide decision making, but discussions of geoengineering will nevertheless be more responsible and more careful if Christian participants commit to treating the world as God's body.

AN INTERCONNECTED, COMMUNAL BODY

But what kind of body is this body of God or body of Christ? By what principles is it governed? A second major idea within the Christian tradition and process thought is the interconnected and communal nature of creation. The good of any individual is connected in a real way to the good of all. Science and theology both affirm this, and much public policy has embraced recycling and other practices which assume a circle of life. Nonetheless, many of our approaches to environmental and social issues still rely on old patterns of thought that do not integrate the idea of interconnection deeply. Christian theology and process thought respond to these by highlighting the pervasive relationality at the foundation of all of reality. Unless human beings embrace the systematic and interrelated nature of reality, geoengineering will, at best, continue business as usual.

From Jesus's teachings forward, Christianity has stressed a countercultural message of communal redemption, inverse hierarchies that flip the powerful and oppressed, and an expansion of our boundaries of accountability to include loved ones as well as strangers and enemies across the globe. This message challenges our contemporary culture's stress on individual happiness or salvation, protection of privilege and profit, and self-interested economics and nationalism. Instead, theology highlights the members of the body of Christ and how each is dependent on the other. Traditional Christian theology also emphasizes God's ultimate vision for the world as one of total healing and salvation. This is evident in Jesus's preaching of the Kingdom of God and subsequent development of this concept in the letters of Paul, the Book of Revelation, and Christian theology. From a different angle, process philosophy also proposes a formative sense of relationality. Whitehead wrote that we are literally composed of our relationships, shaped by all that exists to greater and lesser degrees. These insights concerning relationality from Christianity and process thought seek to describe the nature of our world. These lessons are enormously relevant and important for any discussion of geoengineering because they push us to adopt a new framework of a deep relationality and interconnectivity.

Christian theology claims that the Body of God and the Body of Christ both affirm that God's embodiment is a deeply interconnected organism made up of many distinct and valuable members. In Paul's letter to the Corinthians, he explains this to the early Christians and he emphasizes that all are a part of one body and thus their well-being is interconnected inextricably to other members of the church.

> [20] As it is, there are many members, yet one body. [21] The eye cannot say to the hand, "I have no need of you," nor again the head to the feet, "I have no need of you." [22] On the contrary, the members of the body that seem to be weaker are indispensable, [23] and those members of the body that we think less honorable we clothe with greater honor, and our less respectable members are treated with greater respect; [24] whereas our more respectable members do not need this. But God has so arranged the body, giving the greater honor to the inferior member, [25] that there may be no dissension within the body, but the members may have the same care for one another. [26] If one member suffers, all suffer together with it; if one member is honored, all rejoice together with it. [27] Now you are the body of Christ and individually members of it. [16]

One cannot ignore the suffering or joy of other members, nor can one simply write off one's dependence on the greater whole. Indeed, the apostle Paul emphasizes the importance of the oppressed members of the body and the obligation of Christians to care for one another and share both suffering and joy. There are values and priorities that are the reverse of many of our cultural worship of power and money. The ones who seem weaker are indispensable. The overall health of each member is dependent on the overall health of the entire body. This adds support to Kevin O'Brien's argument in chapter 10 of this volume that decisions about various geoengineering options should preference the input of poor and marginalized people who are most impacted by these possibilities. The Christian tradition theologically and ritually claims Christ's body is alive and active today, present in the struggles, joys, and work of the members of the church. This prioritizes the oppressed and those who suffer, emphasizing the importance of those parts of the body who suffer most.

To see the body of God present in human communities could seem to uphold a very anthropocentric view of the world, which would be troubling to environmentalists. And to say that the body of God is only present in Christian communities would be troubling to anyone who seeks broader inclusivity. Sallie McFague's work helps expand the Body of God to include all the natural world, not just human communities. She discusses how Christianity has been thoroughly human-centered (anthropocentric) but describes this as a weakness of the tradition when it fails to see that God is invested in the redemption of all creation. Humans share a common story of evolution with all of life on the planet, beginning with the big bang. We cannot separate ourselves out from the networks and elements that support our lives. Thus, care for the entire ecosystem becomes essential. In addition, there are grounds in her theology for including all people, not just Christians, within this body as it includes all of creation.

Beyond the Body of God/Christ motif, one finds other theological grounds on which to support our understanding that our individual well-being is connected to a communal vision of restoration. As mentioned earlier, process thought envisions the reign of God as present when the world embodies God's best vision for the world. Ecotheologian H. Paul Santmire also emphasizes a holistic vision of healing which is God's ultimate aim for all of creation. God's vision for the restoration of all of creation is evident in biblical and theological texts. Santmire grounds his own discussion of earth-care in eschatology rather than sacraments or the body of God. In his book *Nature Reborn: The Ecological and Cosmic Promise of Christian Theology*, Santmire explains that the Hebrew Scriptures (Old Testament) and the New Testament both emphasize the final restoration of all of creation, not just humans.[17] Building on scripture, early and classical Christian theologians developed the idea of the kingdom or reign of God to include the natural world because it is part of the Creator's work and intricately connected to our own thriving. With great detail, he chronicles this "cosmic promise" for all. While Christianity has popularly become fixated on the destination of individual souls, Santmire illustrates that this is neither its original nor ultimate vision.

Even if one does not follow Santmire and McFague in extending the Body of Christ or the vision of salvation to all creation, Christians are called to acknowledge that the members of Christ's body are worldwide, both privileged and poor. And Christians are called to care for all other members of the body lest the whole body will fall. Jesus pushes his followers to uncomfortable extensions of concern for others. There are the "greatest commandments" to love God and "love your neighbor as yourself" (Mark 12:29–31, Matthew 22:36–40). Jesus also says that his followers are to love their enemy (Matthew 5:44). Many of Jesus's parables feature a good person who demonstrates what it means to follow God; most of these people violated traditional religious purity codes or were members of a nation or culture that had been a long-time enemy of Jesus's followers. Perhaps the best-known parable like this is that of the Good Samaritan who stops to help a man who has been robbed. Samaritans were foes of the Jewish people, and yet the Samaritan is the example of the good neighbor.

It is not as easy as simply looking out for our own children's futures; Christians are pressed to consider *all* children's futures, loved ones and enemies, here and afar. This is radical and difficult. It calls me to embrace a wider group of accountability than simply myself, my family, or my nation. We are part of an interrelated body, global in scope, and so we are called to love. This involves feeding, healing, and caring for the people themselves and for the planetary system that supports their existence. Rather than a fundamental priority of self-interest, Christian followers of Jesus's way are to be oriented toward God and neighbor.

So, it is fair to say that Christianity is communal and social in its emphasis, pressing us to look beyond ourselves to serve the wider human and natural community. This is against the grain of an individualistic culture that stresses both competition and individual happiness. Since the start of his papacy, Pope Francis has critiqued capitalism's disregard of the poor, and he has been suspected of being socialist or communist for espousing traditional Christian theological commitments to the common good or communal body of Christ. Within the past year, he also issued an encyclical calling for a new age of ecological integrity that honors creation as a sacrament.[18] The Pope also reflects extensively on interconnection and unity within nature. Evoking the work of the saint from whom he took his name, the Pope states that human connection to nature is not simply a romantic idea but rather provides the basis for new action:

> If we approach nature and the environment without this openness to awe and wonder, if we no longer speak the language of fraternity and beauty in our relationship with the world, our attitude will be that of masters, consumers,

> ruthless exploiters, unable to set limits on their immediate needs. By contrast, if we feel intimately united with all that exists, then sobriety and care will well up spontaneously.[19]

From a different perspective, process philosophy also claims that the world is intimately interconnected as a whole. It pioneered a new way of thinking about the relationships between and within all that exists. Whitehead's concept of "internal relations" is his most radical offering to twentieth-century thought. Based on quantum physics' understanding of reality as interconnected energy events, Whitehead affirms that everything is internally related rather than being bits of unconnected matter (or separate persons) that influence one another externally. He claims that we are actually composed of our relationships at every level of our being. We are literally selves-in-relation to the air we breathe, the food we eat, the people we know, our genetic inheritance, the jobs we hold, the modes of thought which predominate, diseases we catch. Even the language we use to think is an inherited symbol system. We are a part of a relational ebb and flow of influences at all times. We do not have a separable existence "external" to our relations to the past, our relationships, environment, culture, genetics, etc.

For example, consider the idea that the water on our planet is a "closed system." This means that the same water has been around since the earth's inception. Our own bodies are up to 60 percent water, and the water that comes into, circulates through, and leaves our body has taken many forms prior to our incorporation of it.[20] The water is part of our organs, our muscles, our blood, and our bones. A U.S. Department of the Interior Geological Survey document prepared for the public explains this interconnection between human beings, history, and other creatures in these playful words: "It is entirely possible that the water you drank for lunch was once used by Mama Alosaurus to give her baby a bath."[21] We are shaped by and influence others perpetually. In an era in which many have grasped the importance of recycling and widespread impact of humans on climate change, we have learned to think more along the lines of Whitehead's insight. But we have not fully integrated this understanding, and this becomes more clear when we consider some key concerns about geoengineering.

GEOENGINEERING AND INTERCONNECTION

Deep relation and interconnection is at the center of our biggest concerns over any global, large-scale environmental manipulation, including geoengineering. While we need to apply our best thinking to creating long-term responses to climate change, we are concerned about setting off negative, environmental chain reactions that further deplete species or harm human communities. We can all think of examples when humans, with the best of intentions, have made disastrous choices and destroyed ecosystems, species, and human cultures. Jared Diamond's best-selling book *Collapse: How Societies Choose to Fail or Succeed*,[22] provides numerous examples of human societies that fell as a result of their own choices. Many of these failures result from our failure to grasp the systematic and interrelated nature of reality.

Process thought addresses this failure in our modes of thought directly. Whitehead contrasted his philosophy of organism, which supposed an interrelated and coconstitutive reality, with what he called a reigning "mechanistic" mode of thought in which we act as though the world is a machine where one can simply fix or replace a part without having an overall impact on the whole. While the scientific revolution brought many indisputable goods, Whitehead wrote about the dangers in approaching the world solely through the perspective of data or isolatable, noncontextualized events, solutions, or facts. In addition, our society makes much ado about individual success and choice without fully recognizing that we are part of a global network of relations where our health and well-being are interdependent with the health and well-being of others. Falling back on old patterns of thought that support ideas such as a "self-made man," we tend to scoff at romanticized examples of interrelation such as "the butterfly effect," which suggests that small-scale environmental changes, such as a butterfly flapping its wings, can create large-scale, systemic impact. It is difficult to conceive that our individual actions have systemic results. Our actions, then, are misguided by partial or short-term ends, and individual good is often conceived outside the realm of common good.

Christian theology has long recognized this deeply-seated tendency of human nature. John Stuart Mill, a founding economist, also did. He proposed that economics could base itself on this foundational principle and built capitalist theory on the assumption that our habits of thought are tuned to self-interest. Christianity, on the other hand, does not see this as a natural human orientation but rather a distortion of our created nature, caused by sin. In original human nature, we would be turned to God, and this would also lead us to be turned to what is good, true, and just for the neighbor and creation. Thus, self-centeredness or being "curved in on oneself" is the predominant definition of sin for Christian theology over the past the two thousand years. It is a tendency to struggle with and ask for human and divine help to mend, not to accept, naturalize, or encourage. This is why Pope Francis's recent environmental encyclical calls crimes against creation a sin fostered by greed and self-absorption.

Contemporary academics also suffer from a lack of integration and relational thinking. John B. Cobb Jr. has also pointed out that the same kind of self-centeredness is evident in the specialization of academic disciplines. That specialization has led to great advancements within each specific field, but we end up with people who only know how to approach issues from within the methods and assumptions of a very limited framework. Thus, academics publish papers no one will read beyond their discipline. Engineers learn a mechanical and scientific approach to problem-solving. Economists within capitalism are pressured to pursue growth at almost any cost. The unintended costs and benefits to the environment and human communities are literally titled "externalities"[23] in capitalist economics. Businesses also generally seek to maximize profit without factoring potential long-term social and environmental impacts. In an interrelated world, we become increasingly siloed and myopic.

When we are dealing with the global issue of climate change, which impacts all things living and nonliving, our solutions need a wide, cross-cultural and interdisciplinary set of perspectives and methods. We cannot hand the reins over to a small group of specialists whose perspectives are shaped by a particular culture or discipline. Rather, we need to embark on what Thomas Bruhn, Stefan Schäfer, and Mark Lawrence have defined as transdisciplinarity in chapter 2 of this volume—that is, research that includes many fields and stakeholders in its deliberations in order to benefit science and society.

Again, philosophical and theological ideas that see reality as fundamentally interconnected may seem like common sense, but several individual or discipline-specific foci prevent us from working productively together to maintain a comprehensive vision of the environmental issues and solutions before us. We are ill-equipped to deal with such global and intricate systems which involve living and nonliving aspects of our earth. We often resort to approaching systematic issues with fixes that presume the world is not internally related. We act as though the world consists of little, separable building blocks. In a discussion of this chapter, a colleague jokingly offered that he perceived the field of geoengineering as a discipline filled with men for whom this is an exciting challenge because it is like building with Legos on a much bigger scale. One can build a machine, replace parts, build it differently, etc. But such engineering only works with earth's ecosystems if you assume the world is bits of matter that do not influence or shape one another. Whitehead called the "Lego" approach mechanistic thinking and proposed a contrary philosophy of organism where the interconnectivity of all events impacts the whole. While Whitehead's philosophy does not argue against geoengineering per se, it would insist that consideration of any ecological solutions must begin and end with examining the irreversible and interrelated systematic changes, at each level, of any approach. When one replaces or influences any aspect of the global environment, there are deep systematic impacts and thus one cannot simply engineer a localized or specific fix.

A philosophy of internal relations would shape how geoengineering is approached. For example, a shaping guideline in current conversations about geoengineering considers it important for solutions to be "reversible" if they cause damage. But in an interconnected world, nothing is ever truly reversible. Each event influences others, setting off a permanent chain reaction of events which incorporate that past influence. There can be greater and lesser influences to be sure, but actions genuinely change future possibilities and outcomes. The state of the past cannot be repeated. Thus, one of the most interesting geoengineering options, the planting of kelp forests to sequester carbon, impacts all the fish and plant species and all aspects of the environment in small and large ways. Adding, removing, or relocating kelp within the ocean changes environs, species, water and food sources. This does not mean we should not do it; since we are always, as said earlier, engineering our surroundings to some degree. But we should act without the idea that we can never simply take any action without creating long-term impacts.[24]

CONCLUSION

This chapter has offered two primary and connected Christian theological and philosophical concepts: the world as God's body and the deep, communal interrelation at the heart of creation. Both of these entail particular claims about the real world, proposing useful values and approaches for considering particular avenues for geoengineering. From a Christian perspective, we are pushed to ask: Who or what will be harmed? How is the common good of all creation impacted and how are the needs of the less powerful prioritized? In what way does a particular geoengineering solution encourage the cooperation and well-being of the whole body of human citizens and creation? If some species or people benefit enormously and profit in the billions while others are displaced from their homes, we cannot call an action just nor can we claim that it is for the good of all. Christians are called, by their faith, to serve the needs of the widest possible "interest group," all living and nonliving creation. Envisioning God's body as really present in the neighbor and all the earth demands the kind of stewardship to bodies described in Hartman's chapter herein. We are not simply scrubbing the air or capturing carbon, we are modifying and impacting the tangible, knowable, living God with whom Christians seek to live faithfully and responsibly. Comfortable parochialisms and pressing engines of self-interest are challenged by basic convictions of the faith. No doubt the pressure for wide-scope solutions will be increasingly high as we suffer various impacts from climate change caused by increased CO_2 levels. Thus Christian theological and philosophical perspectives on communal identity and deep relationality challenge or widen the wisdom of the world's vision of what "successful" solutions entail.

NOTES

1. Sallie McFague, *The Body of God: An Ecological Theology* (Minneapolis: Fortress, 1993).
2. McFague, *The Body of God*, 132.
3. McFague, *The Body of God*, 133.
4. Ibid.
5. All biblical quotes from *The New Oxford Annotated Bible, New Revised Standard Version with the Apocrypha, Third Edition*, ed. Michael Coogan (New York: Oxford, 2001).
6. Larry L. Rasmussen, *Earth-Honoring Faith: Religious Ethics in a New Key* (New York: Oxford University Press, 2013).
7. For further discussion of this, please see Alister E. McGrath, *Reformation Thought: An Introduction, 4th edition* (London: Wiley-Blackwell, 2012), chapter 9.
8. Robert W. Jenson, "Part Two: The Sacraments," in *Church Dogmatics, Vol. 2*, eds. Carl E. Braaten and Robert W. Jensen (Philadelphia: Fortress, 1984), 337.
9. Martin Luther, "Lectures on Genesis," in Volume 1 of *Luther's Works* (55 vols.; Saint Louis: Concordia; Philadelphia: Fortress, 1955–1976), 21–22.
10. Rosemary Radford Ruether, *Gaia and God: An Ecofeminist Theology of Earth Healing* (New York: Harper Collins, 1992), 235.
11. Alfred North Whitehead, *Process and Reality: An Essay in Cosmology*, eds. David Ray Griffin and Donald W. Sherburne (New York: Free Press, 1978 [1929]).
12. Charles Birch and John B. Cobb Jr., *The Liberation of Life: From the Cell to the community* (Cambridge: Cambridge University Press, 1981).
13. Marjorie Suchocki, *The End of Evil: Process Eschatology in Historical Context* (Eugene, OR: Wipf & Stock, 2006).
14. Charles Hartshorne, *The Divine Relativity: A Social Conception of God* (New Haven, CT: Yale University Press, 1948), 89.
15. Ibid., 148.
16. 1 Corinthians 12:20–26.
17. H. Paul Santmire, *Nature Reborn: The Ecological and Cosmic Promise of Christian Theology* (Minneapolis, MN: Fortress, 2000).
18. Francis, "*Laudato Si':* On the Care of our Common Home," 2015, accessed October 10, 2016, http://w2.vatican.va/content/francesco/en/encyclicals/documents/papa-francesco_20150524_enciclica-laudato-si.html, 8. Here Francis is quoting Patriarch Bartholomew (1940–), Archbishop of Constantinople-New Rome and Ecumenical Patriarch (1991–).
19. Francis, *Laudato Si'*, 11.
20. U.S. Geological Survey (USGS) Water Science School, "The Water in You," July 27, 2015. http://water.usgs.gov/edu/propertyyou.html, accessed September 17, 2015.
21. U.S. Geological Survey (USGS) Water Science School, "Where Is Earth's Water Located?" http://water.usgs.gov/edu/pdf/earthwherewater.pdf, accessed September 17, 2015.
22. Jared Diamond, *Collapse: How Societies Choose to Fail or Succeed* (New York: Viking, 2005).
23. This term, "externalities," is applied to social or environmental benefits or harm incurred by economic production. Whether externalities are repaired through taxes or fines is a matter of national and regional governance. The ideal of an "unregulated" market pushes against the inclusion of externalities in determining economic growth or decline; it also pushes against fines and taxes to compensate for harm to human and wider ecological health.
24. I would like to thank Jim Stewart, a geoengineer with Ocean Forresters of Los Angeles, for his presentation on kelp forestation and CO2 sequestration in the geoengineering discussion at the Pando Populus conference in Claremont, CA, June 2015.

BIBLIOGRAPHY

Birch, Charles, and John B. Cobb Jr. *The Liberation of Life: From the Cell to the Community.* Cambridge: Cambridge University Press, 1981.

Diamond, Jared. *Collapse: How Societies Choose to Fail or Succeed.* New York: Viking, 2005.

Pope Francis, "*Laudato Si'*: On the Care of our Common Home." The Vatican. (2015). Accessed October 10, 2016. http://w2.vatican.va/content/francesco/en/encyclicals/documents/papa-francesco_20150524_enciclica-laudato-si.html.

Hartshorne, Charles. *The Divine Relativity: A Social Conception of God.* New Haven, CT: Yale University Press, 1948.

Jensen, Robert W. "Part Two: The Sacraments." In *Church Dogmatics, Vol. 2*, eds. Carl E. Braaten and Robert W. Jensen. Philadelphia: Fortress, 1984.

Luther, Martin. "Lectures on Genesis." In Volume 1 of *Luther's Works* (55 vols.) Saint Louis: Concordia; Philadelphia: Fortress, 1955–1976.

McFague, Sallie. *The Body of God: An Ecological Theology.* Minneapolis: Fortress, 1993.

Radford Ruether, Rosemary. *Gaia and God: An Ecofeminist Theology of Earth Healing.* New York: Harper Collins, 1992.

Rasmussen, Larry. *Earth-Honoring Faith: Religious Ethics in a New Key*. New York: Oxford University Press, 2013.

Santmire, H. Paul. *Nature Reborn: The Ecological and Cosmic Promise of Christian Theology.* Minneapolis, MN: Fortress, 2000.

Suchocki, Marjorie. *The End of Evil: Process Eschatology in Historical Context.* Eugene, OR: Wipf & Stock, 2006.

Whitehead, Alfred North. *Process and Reality: An Essay in Cosmology*. Eds. David Ray Griffin and Donald W. Sherburne. New York: Free Press, 1978 [1929].

Chapter Five

Time's Arrow and Narratives of Climate Engineering

Forrest Clingerman

Climate change is a problem of time. As Saffron J. O'Neill and Nicholas Smith suggest, "The topic of climate change is suffused with the concept of time. Even the term 'climate change' . . . directly references [temporal qualities]: what was climate before; how has it changed; what is climate like at present; and what does it mean for the future?"[1] To this we can add other questions that similarly show the connection between time and climate change: How soon before human culture acts? When will the United States and other countries come up with appropriate plans for mitigation and adaptation? How fast will the oceans rise or ice sheets melt? How many generations of CO_2 remain in the atmosphere? When do we reach a catastrophic tipping point—or is it already too late? Certainly the effects of climate change are frequently portrayed in spatially measured ways: sea level; areas of permafrost and glaciers; species habitats and ranges; and so on. But just as it impacts (and is impacted by) the spatial dimensions of the global environment, the crisis of climate change is channeled by time's arrow.

The unique relationship between climate change and time can be compared to the way other environmental issues have been swayed by temporality. Contemporary environmentalism has been greatly influenced by a portrayal of nature in relation to human history. Thus time becomes reduced to the progression of ideologies within human society. For example, environmental problems stem from the rise of the "modern era," which is characterized by the rise of ideas: industrialization, consumerism, technology, and post-Copernican science. In such situations, "nature" remains as a timeless preserve outside human culture—or else a concept used to critique the problematic culture/nature bifurcation. Climate change, however, attends to time quite differently: it is understood as existing at the nexus of culture and nature, ideology and materiality. On one hand, the temporality of climate change is understood by making normative claims about our evolving conceptualization of the human role in the broader environment, as well as the interweaving of human society and more-than-human physical systems. On the other hand, the temporality of climate change exists beyond ideology through concrete marks made by the ongoing processes of atmospheric chemistry, migration patterns, growing seasons, glacier loss, annual emissions, energy use, and so on. In this unique case, temporality serves as a boundary and meeting point for humans to exist in the material world; time is elemental in turning weather into a human experience of one's climate.

It is necessary to acknowledge that climate change has such a unique temporal dimension because the complexity of time magnifies the spatial changes to nature. This makes such changes more difficult to respond to. Peter Scott writes that when confronted by climate change, "[w]e are being pressed about two scales: the spatial and the temporal."[2] He rightly notes that both of these scales are even more complex than we commonly believe, which means that our response requires a reflection on the biohistorical—that is to say, *bios* and *historia*, material life and culturally demarcated time—intertwining of humanity and nature. At the very least, the interrelationship between life and history forces us to admit that climate change assumes a temporal reference point: a supposed "stability" of the climate that characterizes an Earth than has been hospitable to human existence.

The present chapter claims that theology has an important role in understanding narratives of climate engineering, especially as those narratives deal with issues of time and the human response to climate change. Approaching climate change through the lens of temporality allows us to approach in a novel way some of the interpretive dimensions of climate engineering. The argument will develop in three stages. First, I will examine how contemporary discussions of climate engineering have a poorly defined sense of time, which equivocate between a shallow and a deep sense of temporality. Second, I contend that this equivocation is reproduced in the narratives we use to describe climate engineering. With an appropriate narrative we can uncover a richer temporality to climate engineering, yet there is all too often a danger that our narratives are one-sided. Finally, I will argue that theological reflection is required for understanding the forms that a proper narrative of climate engineering will take. Specifically, theological reflection offers the possibility of transcontextual narratives of climate engineering, focused on mediation and critique of these forms of human-climate interaction.

CLIMATE ENGINEERING AS AN ORIENTATION TOWARD TIME

Acknowledging that climate change is grounded (at least in part) in a sense of time offers a new lens through which to understand our response to the issue. If climate change is time-bound, then our responses also are related to a sense of time. This is certainly the case in mitigation and adaptation: policy analysts and activists debate the speed at which, for instance, emissions reductions must occur or vulnerable populations moved. Time is likewise an element of climate-engineering proposals.

The present section will focus on a seemingly simple question: What is climate engineering, and how does it relate to time? As discussed in the introduction and other chapters of this volume, climate engineering is an attempt to intentionally manage aspects of the climate system, using projects that harness, mimic, or replace natural climate systems. The purpose is not to repristinate nature, but to limit the damaging effects to human society caused by global warming. The term seems to indicate a well-defined, uniform technological intervention into the workings of the Earth's climate. This is incorrect: geoengineering is not a single intervention, but instead a variety of proposals that could potentially be used together, as well as with forms of adaptation and mitigation. The diversity of geoengineering proposals includes such ideas as iron fertilization of the ocean to create algae blooms, introducing particulates into the atmosphere, and launching space mirrors. Even with the differences between carbon dioxide removal (CDR) and solar radiation management (SRM), as well as between what Robert Olson has called "soft" versus "hard" geoengineering,[3] there are nonetheless certain commonalities: geoengineering proposals are humanly designed and implemented projects that attempt to reverse some or all of the effects of climate change on a global scale. Further, such projects define success in relatively anthropocentric ways. This is particularly the case with SRM proposals, which will not resolve some ecological problems associated with increased greenhouse gas concentrations (e.g., ocean acidification) and meanwhile add new ecological damage (e.g., increased "acid rain" and changes to monsoon patterns).

Climate engineering has only recently become worthy of serious reflection and study, taking a place alongside other responses to climate change. Until only recently, political and ethical discussions have focused on two categories of responses to climate change: mitigation (proposals that focus on reducing greenhouse gas emissions) and adaptation (proposals that seek to implement changes to social and individual lifestyles in order to accommodate the environmental effects of climate change). Climate engineering has received serious attention only in light of the global inability to embark on such responses wholeheartedly. Of course there are similarities between mitigation, adaptation, and climate engineering—a fact described in attempts to create a new typology of climate responses, notably the recent ones by Boucher et al. and Heyward.[4] Furthermore, the interconnections between climate engineering, mitigation, and adaptation are relevant to future international and national policy. Discussions of climate change continue to prioritize the need for mitigation and adaptation: the desire to "restore" the "natural climate" promotes mitigation, while the realization that our actions have already altered the climate necessitates adaptation.

But there is nonetheless something unique about climate engineering, as might be inferred from the gradual way the IPCC has considered the role of geoengineering in responding to climate change.[5] In particular, aspects of the intention and scope of climate engineering create points of difference from mitigation and adaptation. Climate engineering recommends a unique conceptual position as part of its political and scientific response: it theorizes a movement from the acknowledgment of past human technological failure to an optimistic solution resting on the increased technological mastery of our global habitat. The presumption that humans successfully can undertake Earth systems management[6] has—until recently—seemed far-fetched and dangerous. Yet this presumption is central to geoengineering, and therefore has been an important aspect of the debate over whether climate engineering is a form of "playing God" or a necessary step toward a new human stewardship of the climate. What is the proper place of climate engineering? Any reply must contextualize the relationship between mitigation, adaptation, and geoengineering. Geoengineering presumes the need for intensive human management before any mitigation is possible. At the same time, climate engineering does not see a sanguine acceptance of our damage to the climate as sufficient to ameliorate the problem at hand.

It is in this context that the question of time becomes more important. In this late hour, how might the temporality of climate change impact an understanding of technologically oriented climate interventions? The temporality of climate change indeed exists as an undercurrent in scientific proposals and ethical analyses of climate engineering. Yet because it is not explicitly described, geoengineering connects with the time of climate change in contradictory ways. To simplify these contradictions, I suggest that climate engineering's relationship with time exists between two levels. On a basic level, climate engineering sees time as an objectively neutral and unchanged substrate—a marker for the chronology of technical manipulation. Thus a superficial level sees time in terms of *chronos*, meaning a view of time as measured sequences of insignificant and interchangeable intervals. Yet on a deeper level, the underlying temporality of climate engineering is based on these technological interventions change our interpretation of human relationship to the climate. That is, climate engineering might encounter time as *kairos*, or viewing the event of climate intervention as the inbreaking of meaningfulness by virtue of being a great moment of history or even "the fulfillment of time."[7]

At its most superficial level, it is possible to reductively define geoengineering technologies as being yet another manipulation of finite and material nature, which exists in the measurement of an anonymous chronology. From this perspective climate engineering technologies—despite any perception of their world-changing possibilities—in truth only exist as attempts to engage a well-defined segment of a linear, uniform timescale of "nature," which remains independent of human culture. In other words, climate engineering takes place *within* a substrate of neutral time, but does not alter it.

From this perspective, climate engineering does not affect the *meaning* of time. Time is seen to continue onward as it has for millennia; the safety of human society requires only an impermanent, momentary, and meaningless shift in the material climate. As Olson writes, proponents " . . . have concluded that geoengineering needs to be studied as a possible 'insurance policy' or 'Plan B' in case we fail to reduce greenhouse gas emissions fast enough to prevent dangerous climate change."[8] Geoengineering is promoted as a *temporary* intervention meant to limit the worst effects of physical climate change over years or decades, while human culture addresses mitigation and adaptation in more *permanent* ways. To some, this means geoengineering serves as a sort of "atmospheric restoration": "The goal is to return the atmosphere to a less degraded or damaged state and ultimately to its preindustrial condition."[9] Such interventions are ephemeral, undesirable, but necessary against the backdrop of the supposed continuity of Nature: in effect, from this perspective *climate engineering is an attempt to stop the clock of the changing climate for a short duration, until society can domesticate the atmosphere such that the climate is aligned with our cultural rhythm.* This intervention admits only a slight degree power over time, but nonetheless presumes that we can domesticate the anonymity of time through the actions of human science and technology. Furthermore, time itself has no real meaning, apart from what is bestowed upon it by human society. To stop the clock of climate change, then, assumes that humanity confronts the moment of our control of the climate as a moment in time like all others. This is not the inauguration of some new human-time, but yet another moment in our ongoing sense of *chronos*.

In this case, what defines the relationship between climate engineering and time? On this level, climate engineering controls the physical changes to the climate without attempting to change the meaning of time's arrow. Time is merely a uniform measurement of humanity's tumultuous relationship with nature. Paul Crutzen wrote in his influential 2006 essay that jumpstarted the discussion of geoengineering, "although by far not the best solution, the usefulness of artificially enhancing earth's albedo and thereby cooling climate by adding sunlight reflecting aerosol in the stratosphere . . . might again be explored and debated as a way to defuse the Catch-22 situation [we are in] . . . and additionally counteracting the climate forcing of growing CO2 emissions."[10] Climate engineering is defined as a necessary but undesirable option in a world that has become defined by risk and equally distasteful choices. Such an option assumes the global environmental changes occur within a chronology bereft of temporal meaning, which requires a separation between temporality, materiality, and climatic meaning. If time is seen in such a linear, uniform, and shallow way, there need not be any significant difference between arguments for research into and implementation of climate engineering. As Keith and Caldeira argue, "[l]ike it or not, a climate emergency is a possibility, and geoengineering could be the only affordable and fast-acting option to avoid a global catastrophe."[11] In statements like this, research is due preparedness. Unless there is a depth to time, research and implementation are equivalent technological projects to control of the temporality of the climate.

But if climate engineering is an attempt to stop and turn back time, how can it be solely a temporary, material act? Or is there a possibility that geoengineering brings with it the inbreaking of a moment of meaning, a new sense of human and natural time? We are now pointed to a second, deeper level in the relationship between time and climate engineering. This level centers on the ways that geoengineering infuses more meaning into time than the arguments of Keith, Caldera, and Crutzen suggest. In contrast to seeing climate engineering as temporary and meaningless material intervention, a second interpretation of the temporality of climate engineering centers on how climate engineering marks a new human time of nature, as well as a new natural time of the human. The events surrounding climate engineering thus occur within special, unique time of destiny and choice, a sort of *kairos* of the atmosphere.

Moving away from a presumption of the superficiality of time, what happens when time is seen as a vehicle that participates in the construction of meaning? Climate engineering technologies, from this point of view, disrupt the natural sequence of time—and the human relationship to climate stability and change—to inaugurate a time of human atmospheric guidance. In turn, geoengineering becomes the name of a radical temporal transition: the end of one climate and the beginning of another. In other words, climate engineering names a temporal boundary, which demarcates a new human bond with the atmosphere across a geological timescale.

Such a transformation of temporal meaning is not unequivocal. Rather, within this deep sense of time there exists the possibility of either dread or hope. For example, the seriousness in which climate engineering is considered grows in connection with an increased pessimism toward global and national climate policy mechanisms offering resolution through mitigation and adaptation—increased "climate pessimism" has equaled increased calls for studying climate engineering. Climate pessimism is, itself, a response to this timely matter: we are running out of time, and so the urgency of climate change calls for more drastic measures. In contrast, some advocates see climate engineering as part of a chastened human footprint—an opportunity to find a better environmental balance. Christopher Preston notes this possibility in his discussion of the presumptive argument against climate engineering. Rather than always being about human pride, geoengineering "could be a serious attempt to make amends. If the intention is to lessen human suffering, to protect non-human species, and to preserve environmental values, then it is not clear that geoengineering adds to the blight."[12] Similarly, Kevin O'Brien writes how climate engineering might be undertaken without blind hubris when undertaken as part of the act of repentance.[13] O'Brien's view is a bit more suspicious of geoengineering than Preston's, yet both authors see ways in which it is possible for climate engineering to not be marked by hubris. Such attempts signal a more hopeful technological relationship between human and climate.

A deeper perspective of time—one that suggests that the temporality of climate engineering includes a *kairos,* a unique moment—creates a new interpretation for the human relationship to the atmosphere and material world. In effect, from a second perspective climate engineering is an attempt to usher in a new period. *In other words, climate engineering domesticates the temporality of the climate itself, in order for climate change to be reordered as a cultural relationship to the atmosphere*. We are no longer beholden to time, but are masters of it. Unlike the first level, more is at stake than simply deciding whether climate engineering is an appropriate material solution for an (albeit complex) problem. Rather, it becomes a proposal to reconceptualize the temporal ebb and flow of how humanity lives on the planet—to rework time—in order to interrogate how the wickedness of the problem of climate change is an encounter with the roots of human experience. Such atmospheric domestication reinterprets the *meaning* of time, in ways that force us to see climate engineering as more than merely a technical solution to the problem of climate change.

What I've suggested thus far is that climate engineering proposals are caught between two modes of time. In some senses, climate engineering seems to be developing as a technical solution bereft of temporal meaning: it concentrates on modeling Earth systems without concern for the deeper impacts its changes bring. But in other senses, climate engineering proposals seem to understand how our domestication of the climate will inaugurate a new interpretation of the relationship of human society, nature, and time. In other words, the debate over climate engineering does not presume a superficial chronology, nor a deep sense of *kairos*. Instead, climate engineering is predicated on the ambivalence between the two.

This tension is not limited to climate engineering—it is also apparent in discussions of the Anthropocene. After all, climate engineering is climatic parallel to the geological debate over the Anthropocene epoch. Originally suggested by Crutzen and Stoermer, the Anthropocene can be described in narrow or broad terms, akin to the two temporal interpretations of climate engineering we are discussing. In a narrower sense, the Anthropocene is a proposal that the Earth has entered into a new geologic time period, which is characterized by profound and pervasive human impacts. Geologists debate the material markers and dating of this period, with an eye for understanding the "Critical Zone," "the surface or near-surface environment where the geosphere, pedosphere, hydrosphere, biosphere, atmosphere, human beings, and the energy from the Sun all interact."[14] Such a view sees time in disinterested ways. But there are also ethical and cultural implications to understanding our changing relationship to the geologic record. Therefore in a broader sense, the term "Anthropocene" has also taken on a cultural meaning of acknowledging how humans are now one of the primary drivers of environmental change. This makes the question about more than geology—it is also about how we should define our relationship with nature, and how we now have broken into a new age. And so, " . . . unlike other geological time unit designations, [how we define the Anthropocene] will probably have effects beyond geology. For example, defining an early start date may, in political terms, 'normalize' global environmental change. Meanwhile, agreeing to a later start date related to the Industrial Revolution may, for example, be used to assign historical responsibility for carbon dioxide emissions."[15] From this deeper perspective, then, the Anthropocene highlights ontological changes in the human relationship with the planet.

Making an analogy with the Anthropocene, climate engineering is not just a technical solution to climate change, but the change of meaning that has been named the Anthropocene: the event of a novel, technologically motivated hermeneutics of the climate. In other words, climate engineering is one manifestation of the "great moment" of the Anthropocene, insofar as it is a domestication that occurs by transforming and domesticating the materiality of time. When defined as a "new" time, the Anthropocene marks a new human relationship.

Interestingly, however, the Anthropocene itself has a murky sense of time. In some ways it seems clear that the Anthropocene represents a unique moment of time. But it is evidenced through recourse to the objectivity of chronological time. In turn, in light of its superficial chronology, announcing that the Anthropocene is a geological period akin to all others risks undercutting its import as a temporally significant and meaningful. More worryingly, once the Anthropocene is seen as yet another moment in the ongoing materiality of the planet, devoid of a uniqueness that is bracketed by the methods of the natural sciences, we are tempted to equate the Anthropocene to the other epochs of human-environment interaction. But suddenly, this means that as humans living in the midst of nature's chronology, we have always lived in the Anthropocene—a statement that makes the Anthropocene fundamentally meaningless. Is the Anthropocene a moment of meaning, or not?

In such murkiness, we are unable to discern an appropriate human response to climate engineering. What time do we find ourselves in? Perhaps we are caught in a never-ending chronology of human conquest, wherein each moment is finally indistinguishable from others. Discussing the psychological costs associated with climate engineering, Gareth Davies writes that "[g]eoengineering . . . introduces the possibility of purgatory, and of a future of endless managed risk, of a fear that never quite goes away."[16] Similarly Lovelock has remarked, "Are we sufficiently talented to take on what might become the onerous permanent task of keeping the Earth in homeostasis? . . . We could find ourselves enslaved in a Kafka-like world from which there is no escape."[17]

Or perhaps we find ourselves in a more fecund moment of time, a *kairos* of the atmosphere. Scholars who see climate engineering as offering a more benign possibility for humankind lead us in this direction. Instead of the dire prognostication of Lovelock, climate engineering might be part of a better stewardship of the Earth; it can contribute to what Steffen, Crutzen, and MacNeill suggest is the third phase of the Anthropocene.[18] Crutzen writes that "hopefully, in the future, the '*anthropocene*' will not only be characterized by continued human plundering of Earth's resources and dumping of excessive amounts of waste products in the environment, but also by vastly improved technology and management, wise use of Earth's resources, control of human and domestic animal population, and overall careful manipulation and restoration of the natural environment."[19] Steffen et al. are skeptical that climate engineering can ultimately fulfill such a role, but "ecomodernists" are less distrustful, arguing that "knowledge and technology, applied with wisdom, might allow for a good, or even great, Anthropocene."[20] Such technological optimism sees the interpretation of climate engineering as a symbol of the new human relationship with nature, even while others see ambivalence or cautious resignation.

NARRATIVES OF A CHANGING CLIMATE

Climate engineering is caught somewhere between shallow and deep senses of time, between seeing climate engineering in terms of *chronos* and *kairos*. How can we resolve this? Overcoming this conflict means acknowledging that climate engineering is not merely a scientific issue, but just as importantly something that is affected by the narratives we construct about it. One of the strengths of using stories[21] as a means of understanding is that narratives weave together otherwise discordant elements without attempting to reduce the differences between them. The plot of a story is able to gather together discordant senses of time into a single, meaningful whole.[22] Furthermore, narratives can serve as a frame, a tool for understanding. Alexa Spence and Nick Pidgeon helpfully describe the utility of frames: "[a] frame allows complex issues to be pared down and for some aspects of that issue to be given greater emphasis than others in order that particular audiences can rapidly identify why an issue may be relevant to them. . . . Frames can also serve ideological and governance purposes, and as a consequence different social actors within the climate change policy domain (as within other fields) constantly compete in order to present and legitimize their own interpretations of the salient issues."[23] Because of its ability to gather disparate events into a meaningful whole, a story can act as a framing structure that easily assimilates facts and values into our understanding. If narratives are useful as coherent framing devices, it should be apparent why the temporality of climate engineering leads to narratives. The reason that climate engineering finds itself in the midst of different senses of time is the lack of a coherent narrative that interweaves the *chronos* and *kairos* of climate change. Indeed, the need for narrative is part of the reason why climate engineering requires interdisciplinary study: scholars in ethics, philosophy, and religion try to uncover new narratives and metaphors that promote a better understanding of how climate engineering will interconnect human society and global ecosystems.

So if the issue is time, why look to stories? Because there is a close connection between time and narrative in lived experience. Elsewhere in this volume, others have investigated the ways that stories (and related things like metaphors and models) connect to our experiences: for example, Laura Hartman discusses the potential of a medical model of climate engineering, and Willis Jenkins explores large-scale stories. Furthering the connection between climate engineering and stories, I wish to show the connection between such stories and our sense of time. Following Paul Ricoeur, I understand "temporality to be that structure of existence that reaches language in narrativity and narrativity to be the language structure that has temporality as its ultimate referent."[24] In other words, from the lens of narrative, time is not simply a mechanism for objective demarcation, but something that exists as part of lived experience which can only be expressed through the language of story. But at the same time, our stories are ways of expressing our experience of time. Taking up a point expressed by Ricoeur, Hans-Georg Gadamer, and others, temporality is expressed and framed through narrative, while narratives interpret and suggest the meaning of time. Further nuance to this connection is found by acknowledging the interconnection between metaphor and narrative—metaphors are akin to little stories, functioning as shorthand ways of understanding. Both narratives and metaphors are the instruments for describing personal and communal experience, and thereby actualizing temporality for persons and society.

The relationship between time and narrative means the temporal tensions discussed above can in truth be related to uncertainties of our narratives of human interventions in the climate. To show this, we can start with climate change. How is climate change a problem of narrative? In several works, Mike Hulme has suggested that we find ourselves caught between the material and conceptual interpretations of climate change:

> Humanity is firmly embedded *within* the physical climate system, straining further the artificially constructed boundaries between the natural and the human. But at the same time as physical climates are changing, the imaginative *idea* of anthropogenic climate change is penetrating and changing political discourse and social organization in novel ways. The past (through historic emissions of greenhouse gases) and the future (through lurid descriptions of the effects of predicted climates to come) are interacting in new ways to provide a novel motor for cultural change.[25]

In other words, climate change is not merely an abstract phenomenon, but a way of narrating experiences of time. Elsewhere Hulme suggests different narratives of climate change: the Edenic, the Apocalyptic, the Promethean, and the Themisian.[26] For Hulme, these myths show how the idea of climate change works in society. I would argue that they also show the ways that temporality is framed and contextualized through a narrative structure of past, present, and future. The "imaginative idea" of climate change becomes our stories of the complex dynamics, relationships, and materiality of the climate system; the result is a need for reflecting on changing narratives and metaphors we use to understand the climate.[27] At the same time, the diversity of narratives implies diverse structures of time—a diversity that encourages contradiction and ambiguity.

In similar ways, the temporality of climate engineering also is a question of narrative—but which time? Which narrative? Certainly there are several metaphors of climate engineering, as suggested by Laura Hartman in chapter 7 of this volume. More extensive than metaphors are narratives that bind the understanding of climate engineering together, yet no clear narrative has emerged as the orienting one. Thus the uncertainty found in the connection of time and climate engineering, which was described in the last section, is somehow connected with the uncertainty and diversity of narratives in the current debates over climate engineering. In the case of climate engineering (just like in the case of climate change and other so-called "wicked" problems), societal decisions are informed by which narratives and metaphors illuminate our understanding. But as of yet, no clear narrative predominates.

> [Climate engineering] may appear to be a purely technological problem. However, it is also, and perhaps even more so, a societal and an ethical issue. How it is metaphorically, argumentatively and even visually framed can have deep ethical implications which need to be scrutinized. . . . In some respects metaphor [and, I am arguing, narratives more generally], a tool we use to think and act with, is a linguistic technology that needs as much ethical oversight as the technologies we 'see through' it, such as geoengineering. Metaphors are the mind's eyes and society's tools. They provide us with visions of the world and instruments to change it.[28]

As a result, the novelty of climate engineering is forcing us toward new stories and metaphors of our relationship with time.

But we don't know what story to choose. This is perhaps the most important dilemma in the debate over climate engineering, insofar as "the primary purpose of environmental rhetoric has always been to change the future rather than merely to predict it."[29] Just as there are different senses of time, so too do we see many scholars attempting to parse the diverse narratives that are emerging. For example, Paula Curvelo and Angela Guimarães Pereira posit several "master narratives" used to discuss climate engineering in public discussions:

- Narratives of progress and the risk of dismissing a promising technology.
- The narrative of a singular hierarchy of knowledge and the new credo of evidence-based decision making.
- The value science narrative, which requires asking the right questions rather than only providing correct answers to whatever questions are asked.
- The narrative of greater public openness and involvement—upstream public engagement.
- The narrative of no time to lose—the speed imperative, to act immediately, before it is too late.
- The narrative of ethics—the forms, functions, and proper regulation of science through ethics.
- The narrative of failure, which spotlights the side effects and unintended consequences of geoengineering technologies. [30]

The narratives suggested by Curvelo and Pereira are conceptual and contradictory. Some of these narratives promote climate engineering as a "temporary" intervention into an ongoing human time, while others exhibit a more pervasive, radical temporal change. Other scholars have suggested more poetic and metaphorical narratives; Clive Hamilton, for example, has suggested that current discussions are focused on two narratives of climate engineering: Promethean (and its focus on human power and ingenuity) and the Soterian (with a concern for preservation and humility). Others have offered metaphors related to medicine, the idea of a technofix, and "Plan B." There are tensions between these narratives, especially in light of the ways that these narratives frame our understanding. This suggests that we do not have a firm grasp of the meaning of climate engineering, and our narratives are only providing a limited view.

This diversity of narratives points to two things, which are necessary but difficult to hold in balance. First, the discussion of climate engineering has not settled upon a universally acknowledged or accepted narrative—no one has found "the" narrative that must be used. Indeed, as is the case with metaphor, there is generally not a single story that encapsulates the entirety of meaning. Stories are partial, fragmentary, and limited by perspective, which means there is a danger in assigning a single story an exclusive role in the construction of meaning. As seen in the work by Curvelo and Periera, Hamilton, and Hulme, climate engineering requires complementary narrative for understanding. But as of now there is still a lack of narrative cohesiveness in our understanding of climate engineering, as evidenced by the relationship between time and climate engineering. And so, second, the need for a compelling narrative has been left unfulfilled. This is a significant problem, not only for understanding, but also for deciding upon a societal response. Willis Jenkins noted that "[e]thics seems overwhelmed by climate change. None of our inherited moral traditions anticipate practical responsibilities for managing the sky, nor construct institutions of justice to discipline power across cultures and generations, nor imagine harming and loving neighbors through diffuse ecological flows."[31] This dynamic becomes evident in discussions over climate engineering in no small part because we lack a vocabulary and interpretive structure for understanding—precisely what stories offer.

THEOLOGY, NARRATIVE, AND A *KAIROS* OF THE INTEGRITY OF LIFE

Stories help us to understand the meaning of climate engineering. Yet the narratives that we have identified thus far are incomplete and problematic in at least two respects: singly, each one neglects the contradictory timescales of climate engineering, instead favoring one or the other temporal level; and collectively, they lack coherence, thus making it difficult to put forward an interpretation of climate engineering that productively accepts the diverse interpretations found in these stories. Put simply, there is not currently a bridge that brings these narratives together, and yet none is sufficient on its own. Without the capacity to challenge one-sided representations of time, the narratives and metaphors of climate engineering individually remain artificial and one-sided. Overcoming such one-sided representations requires a narrative that faithfully represents the depth of existence and meaning. What is necessary, therefore, is a better narrative. Building such a narrative is something that is gradual, communal, and organic—and thus the aim of this chapter is not to propose the *content* of a better narrative. Instead, the conclusion of this chapter will concentrate on the optimal *form* of this narrative: a transcontextual, theological narrative of the time of climate engineering.

A better narrative for time and climate engineering is one that allows for the mediation of diverse interests and times, which means a narrative that places the possibility of advancing the prospect of "the integrity of life before God." [32] David E. Klemm and William Schweiker suggest this phrase, due to the fact that humans are mediating creatures, caught between competing needs, interests, and interpretations of the world. They note that "[t]he human struggle for wholeness, for integrity, situates us in the wider complex of life as well as amid conflicts among our desires."[33] This not only suggests a need to move beyond our individual self-interest, but just as importantly opens us to the need for transcendence in the midst of a sense of responsibility for the other. The integrity of life—the flourishing of human and nonhuman, individual and collective beings on planet Earth—is at the heart of the meaning of being, and it is only before a sense of the Transcendent that such integrity is accomplished.

More to the point, then, successfully narrating the meaning of climate engineering involves a theological task: the connection between time and climate engineering must become more coherent in conversation with our "ultimate concern," a term Paul Tillich used to identify what we (individually and collectively) consider properly ultimate in light of the meaning of being. One's ultimate concern is ultimate in terms of excluding "all other concerns from ultimate significance,"[34] as well as being unconditional, total, and infinite.[35] At the same time, insofar as it is a "concern," it is something that engages us as an existential, lived aspect of our religious experience.[36]

In light of faith as ultimate concern, Tillich suggested a relationship between *chronos* and *kairos*. Tillich notes for Christianity there is a singular moment in the center of history of the "great Kairos," but beyond this there is the potential for numerous "relative" *kairoi* that are in relation to the power and meaning of this source. Rather than seeing the two as either/or, therefore, the relationship is more dialectical. It is from within our experience of *chronos* that moments of meaning, *kairos*, emerge. Such moments are intersubjective and existential: not only is there an objective element of history present, but at the same time there is an individual and communal recognition of the uniqueness of that moment and its effects. Insofar as these relative moments upbuild and draw upon the Great Moment—a moment that is felt within our experience of ultimate concern—they suggest relative *kairos*, which suggest pivotal moments that shape our sense of being. But, Tillich admonishes, there are two dangers: "first, [*kairoi*] can be demonically distorted, and second, they can be erroneous. . . . The error lies not in the *kairos*-

quality of the situation but rather in the judgment about its character in terms of physical time, space, and causality, and also in terms of human reaction and unknown elements in the historical constellation."[37] In light of this description of time as *kairos*, then, the theological task is uncovering narratives that allow us to envision time correctly, while safeguarding us from distorted or erroneous visions of time. In the case of climate engineering, these distortions can be said to overemphasize human value and power while also underemphasizing values and meanings that transcend the individual human here-and-now.

Seeking narratives of the temporality of climate engineering within the context of theological reflection, then, starts by acknowledging the limitations of current narratives of climate engineering: each addresses a specific point of view, but does not transcend it. More to the point, individual narratives fall prey into defining time in a chronological sense, or in a distorted view of meaning. Looking carefully at the collection of narratives and metaphors discussed above, one can see a continuum that correlates with the uncertainty over the time of climate engineering.

Analogous to this sense of the temporality of climate engineering, two rough groupings of metaphors and stories emerge. One grouping concentrates on a chastened humanity "fixing" the planet (e.g., Curvelo and Pereira's narrative of progress, or Hamilton's Promethean narrative); this, we might argue, is tied to a chronological, objective sense of time. Such senses of time do not include a vision of one's ultimate concern, but remain fixed without a real sense of transcending meaning or value. These stories *domesticate time* and the climate by assuming nature to be an engineering problem rather than something that overflows beyond human control or reason.

The other grouping of narratives (e.g., the value science and Soterian narratives) contains stories and metaphors that attempt to discern how climate change alters the meaning of time. In such stories, there is a possibility for error and distortion. In attempting to uncover the meaning of time, there might be a sense of pessimism that emphasizes what is taken to be human overreach and hubris, or a sense of guarded optimism related to the possibility of a new ecozoic era. In both cases, the narratives are attempts to communicate a value-laden cleavage of the human timescale. This is where the potential for distortion emerges: in spite of their differences, these narratives equally become distorted views that presuppose that humans are the sole authors of global meaning and value. These groupings of narratives do not transcend the limits of an imperfect timescale through which to characterize our relation with the climate. In other words, these narratives cannot gather together the mundane and the ultimate as simultaneously necessary for expressing the desire to see how our ultimate concern frames our relationship to the climate around us.

While a theological narrative has been suggested, a further narrative quality is necessary to overcome the dangers of domesticated or distorted senses of time. There is a need for a *reflexive* narrative, which includes not only both forms of time, but equally the tension between them—a narrative that becomes "transcontextual." Verna Marina Ehret suggests three possibilities of narrative related to religion: "The first is metanarrative, where a particular narrative is seen as both universal and true in an absolute way. The second is the contextual narrative, which is the recognition of the dependency of any narrative on its context, the community, and individuals from which the narrative arises."[38] These two possibilities are incomplete and prone to error. In the case of climate engineering, both contextual and metanarratives are present. Some discussions lend themselves to anonymous, abstract chronologies limited to scientific explanation; this results in mere contextual narratives that presume the realm of climate engineering to be science and engineering. These contextual narratives, furthermore, end up falling prey to the problems of chronological time without meaning. In contrast, other discussions point toward grander narratives of human being, positing climate engineering as a moment akin to Prometheus bringing fire to humanity. Such narratives threaten to become metanarratives that concentrate our vision of meaning onto a few select human and planetary qualities. Such narratives find meaning and value in time, but due to the flaws of metanarrative, they are distorted interpretations of the meaningfulness of time.

What is needed, then, is what Ehret calls a "transcontextual" narrative, or a narrative that offers a satisfying mediation of meta- and contextual-narratives. "This is a blend of narrative frames, bringing together multiple perspectives in order to exist simultaneously in the tension of multiple frames. A trans-contextual narrative . . . is a mediating narrative, which recognizes both the holistic claim of the meta-narrative and the particularity of emplacement found in contextual-narrative."[39] In other words, such a narrative acknowledges the partial nature of our views, while also suggesting a way to bring coherence to our competing visions, values, and interpretations.

Properly understood, a transcontextual narrative of geoengineering serves to reorient our understanding away from just the climate, toward our recognition that any story of climate engineering is a story about the human relationship to the climate. This is a story wherein humans are not only authors but actors, and likewise the atmosphere is not mere backdrop, but creative power. In the case of climate change, then, such a transcontextual narrative becomes a theological vision of hope in the midst of human fallibility, akin to Martin Luther's claim that we are simultaneously sinner and justified. A similar dynamic appears when discussing climate engineering: humanity is confronted by its own past flaws, and left to ascertain whether the very qualities of our past failures will lead to salvation or destruction.

Such a transcontextual narrative has direct relevance to time. A transcontextual narrative of climate engineering suggests that the time of climate engineering is neither shallow nor deep. It acknowledges and accepts our inevitable domestication and overhumanization of time as a manifestation of how we overlook the transcending qualities that time's flow gives to human being and the flourishing of existence. For climate engineering, this means focusing on change and duration without meaning as necessary to explain the functional issues surrounding climate engineering proposals. But through a transcontextual narrative we would also be opened to the possibility of ways that time mediates transcendence. Seeing that the event of climate engineering is an act of meaning, which might be distorting the possibility of human flourishing or advancing it. A transcontextual narrative, in other words, is the story of climate engineering as a moment of choice, a point of reflexive possibility: can we accept the intertwining of value-free chronology as the inevitable context of the potential *kairos* that we are invited to?

A strength of a transcontextual narrative of climate engineering is how it allows an avenue for critique and mediation by first recognizing that climate engineering is just as much a human interpretation of the climate as it is a collection of technologies. Through a desire to find the "integrity of life before God," theological reflection can provide a means to critique the metaphors that domesticate the climate through our technological reduction of timescale. The integrity of life is about mediation and wholeness, extending beyond the shallow limits of our thinking. Ultimately, it seeks the sacred depth of being, which critiques the all-too-human attempts at intellectual and physical mastery.

This critique acknowledges the limits of our metaphors and the boundaries of our sense of time. As Tillich suggests, from a theological perspective the meaning of our temporal experience is oriented toward the "depth dimension," which includes an eternity that contains and transcends time. The temporal narratives of climate engineering, in contrast, reside exclusively in shallow time, bereft of a more fecund meaning. Another way to say this is that both narratives fail to recognize the truth of the other, and therefore neglect the possibility of transcendence. Time *is* an objective flow, and its continuity is beyond human intention or control. Time also *is* filled with meaning and purpose, opening the possibility of a new lived experience. And, finally, time is something more than these, with an overflowing meaning beyond human control or ambition. To ask the question "When exactly did humans attain dominance of the earth's environment?"[40] neglects important prior questions: why do we interpret our experience through terms of mastery and domination, and why have we foreclosed the possibility of temporal wildness?

At the same time, theological reflection is a way to critique the false hope of climate engineering by uncovering narratives of eschatological hope, suggesting that climate engineering is permissible only if it transcends and redeems time. Ultimately, religion raises the specter of seeing climate engineering and climate change in eschatological terms, and judging them accordingly. Stefan Skrimshire has suggested that climate change has caused two shifts in our environmental consciousness. The first is that "humanity has become an agent in the constitution of its own habitat."[41] More importantly for the present circumstances, the second shift sees "climate" as a moral concern, which serves as a medium for our ability to cause harm to others (past and present), and to be harmed. This means that global warming is "problematic to cultural and theological narratives about the 'trajectory' of the human."[42] The story of "climate apocalypse" comes to the fore, both in our discussions of climate change and in our dreams of a technological fix. But is this apocalypse filled with hope or despair? Insofar as climate engineering remains unable to offer a narrative that encompasses both hope and despair, it cannot be an adequate representation of the fecundity of time. The tensions seen in the temporality of climate engineering are echoed in the tensions between narratives, even until "the end."

NOTES

1. Saffron J. O'Neill and Nicholas Smith, "Climate Change and Visual Imagery," *WIREs Climate Change* 5 (2014): 83.
2. Peter M. Scott, "Humanity," in *Systematic Theology and Climate Change*, ed. Michael Northcott and Peter M. Scott (New York: Routledge, 2014), 113.
3. Robert L. Olson, "Soft Geoengineering: A Gentler Approach to Addressing Climate Change," *Environment* 54, no. 5 (September/October 2012): 29–39.
4. Olivier Boucher, Piers M. Forster, Nicholas Gruber, Minh Ha-Duong, Mark G. Lawrence, Timothy Lenton, Achim Maas, and Naomi Vaughan, "Rethinking Climate Engineering Categorization in the Context of Climate Change Mitigation and Adaptation," *WIREs Climate Change* 5 (2014): 23–35; Clare Heyward, "Situating and Abandoning Geoengineering: A Typology of Five Reponses to Dangerous Climate Change," *PS: Political Science and Politics* 46 (2013): 23–27.
5. A. Petersen, "The Emergence of the Geoengineering Debate within the IPCC," Case Study, Geoengineering Our Climate Working Paper and Opinion Article Series, accessed January 18, 2016, http://wp.me/p2zsRk-bp.
6. Brad Allenby, "Earth Systems Engineering and Management: A Manifesto," *Environmental Science and Technology* 41, no. 23 (December 1, 2007): 7960–65.
7. Paul Tillich, *Systematic Theology, Volume Three* (Chicago: University of Chicago Press, 1963), 369.
8. Olson, "Soft Geoengineering," 29.
9. Robert B. Jackson and James Salzman, "Pursuing Geoengineering for Atmospheric Restoration," *Issues in Science and Technology* 26 (Summer 2010): 72.
10. Paul J. Crutzen, "Albedo Enhancement by Stratospheric Sulfur Injections: A Contribution to Resolve a Policy Dilemma?" *Climatic Change* 77 (2006): 212.
11. Ken Caldeira and David W. Keith, "The Need for Climate Engineering Research," *Issues in Science and Technology* 27 (Fall 2010): 57.
12. Christopher Preston, "Re-Thinking the Unthinkable: Environmental Ethics and the Presumptive Argument against Geoengineering," *Environmental Values* 20 (2011): 470.
13. Kevin J. O'Brien, "First Be Reconciled: The Priority of Repentance in the Climate Engineering Debate," chapter 10 of this volume.

14. Matt Edgeworth, Dan deB Richter, Colin Waters, Peter Haff, Cath Neal, and Simon James Price, "Diachronous Beginnings of the Anthropocene: The Lower Bounding Surface of Anthropogenic Deposits," *The Anthropocene Review* 2 (2015): 48.
15. Simon L. Lewis and Mark A. Maslin, "Defining the Anthropocene," *Nature* 519, no 7542 (March 12, 2015): 171.
16. Gareth Davies, "The Psychological Costs of Geoengineering: Why It May Be Hard to Accept Even If It Works," in *Climate Change Geoengineering*, ed. Wil CG Burns and Andrew L. Strauss (Cambridge: Cambridge University Press, 2013): 73–74.
17. James Lovelock, "A Geophysicist's Thoughts on Geoengineering," *Philosophical Transactions of the Royal Society* A 366 (2008): 3888.
18. Will Steffen, Paul J. Crutzen, and John R. McNeill, "The Anthropocene: Are Humans Now Overwhelming the Great Forces of Nature?" *Ambio* 36, no. 8 (December 2007): 614–621. The authors themselves are skeptical of this connection with geoengineering, however.
19. Paul J. Crutzen, "The 'Anthropocene,'" *Journal de Physique* IV 12 (2002): Pr10–4.
20. "An Ecomodernist Agenda," accessed January 18, 2016, http://www.ecomodernism.org/manifesto-english/.
21. I will use "story" and "narrative" interchangeably, although there are differences that might be useful to investigate in another setting. It would also be important to clarify the connection between narratives, metaphors, and models.
22. See the discussion of plot in Paul Ricoeur, *Time and Narrative*, Volume 1 (Chicago: University of Chicago Press, 1983).
23. Alexa Spence and Nick Pidgeon, "Framing and Communicating Climate Change: The Effects of Distance and Outcome Frame Manipulations," *Global Environmental Change* 20 (2010): 657.
24. Paul Ricoeur, "Narrative Time," *Critical Inquiry* 7 (1980): 169. Ricoeur explores this dynamic at length in the magisterial three volume *Time and Narrative* (Chicago: University of Chicago Press, 1984–1988).
25. Mike Hulme, "Four Meanings of Climate Change," in *Future Ethics*, ed. Stefan Skrimshire (New York: Continuum, 2010), 38.
26. Mike Hulme, "Climate Change: No Eden, No Apocalypse," in *Exploring Climate Change through Science and in Society* (New York: Earthscan/Routledge, 2013), 204–6.
27. For a historical overview of how our stories of climate have changed, see Lucian Boia, *The Weather in the Imagination* (London: Reaktion, 2005).
28. Brigitte Nerlich and Rusi Jaspal, "Metaphors We Die By? Geoengineering, Metaphors, and the Argument from Catastrophe," *Metaphor and Symbol* 27 (2012): 133.
29. Hulme, "Four Meanings of Climate Change," 46.
30. These are taken verbatim from Paula Curvelo and Angela Guimarães Pereira, "Geoengineering: Reflections on Current Debates," *International Journal of Science in Society* 4 (2013): 13.
31. Willis Jenkins, *The Future of Ethics: Sustainability, Social Justice, and Religious Creativity* (Washington: Georgetown University Press, 2013), 17.
32. David E. Klemm and William Schweiker, *Religion and the Human Future: An Essay on Theological Humanism* (Malden, MA: Wiley-Blackwell, 2008), 5–6 and elsewhere.
33. Ibid., 35.
34. Paul Tillich, *Systematic Theology*, Volume 1 (Chicago: University of Chicago Press, 1951), 11.
35. Ibid., 12.
36. Ibid.
37. Paul Tillich, *Systematic Theology*, Volume 3 (Chicago: University of Chicago Press, 1963), 371.
38. Verna Marina Ehret, "Globalization, Redemption, and the Dialectics of Courage," *Sophia* 53 (2014): 69.
39. Forrest Clingerman and Verna Marina Ehret, "Hope and Fear: The Theological Side of Framing Climate Change," *Ethics, Policy and Environment* 16 (2013): 154.
40. Bruce D. Smith and Melinda A. Zeder, "The Onset of the Anthropocene," *Anthropocene* 4 (2013): 8.

41. Stefan Skrimshire, "Eschatology," in *Systematic Theology and Climate Change* (New York: Routledge, 2014), 159.
42. Ibid.

BIBLIOGRAPHY

Allenby, Brad. "Earth Systems Engineering and Management: A Manifesto." *Environmental Science and Technology* 41, no. 23 (December 1, 2007): 7960–65.

Boia, Lucian. *The Weather in the Imagination.* London: Reaktion, 2005.

Boucher, Olivier, Piers M. Forster, Nicholas Gruber, Minh Ha-Duong, Mark G. Lawrence, Timothy Lenton, Achim Maas, and Naomi Vaughan. "Rethinking Climate Engineering Categorization in the Context of Climate Change Mitigation and Adaptation." *WIREs Climate Change* 5 (2014): 23–35.

Caldeira, Ken, and David W. Keith. "The Need for Climate Engineering Research." *Issues in Science and Technology* 27 (Fall 2010): 57–62.

Clingerman, Forrest, and Verna Marina Ehret. "Hope and Fear: The Theological Side of Framing Climate Change." *Ethics, Policy and Environment* 16 (2013): 152–55.

Crutzen, Paul J. "Albedo Enhancement by Stratospheric Sulfur Injections: A Contribution to Resolve a Policy Dilemma?" *Climatic Change* 77 (2006):211–19.

———. "The 'Anthropocene.'" *Journal de Physique* IV 12 (2002): Pr10–1 to –5.

Curvelo, Paula, and Angela Guimarães Pereira. "Geoengineering: Reflections on Current Debates." *International Journal of Science in Society* 4 (2013): 1–21.

Davies, Gareth. "The Psychological Costs of Geoengineering: Why It May Be Hard to Accept Even If It Works." In *Climate Change Geoengineering*, edited by Wil CG Burns and Andrew L. Strauss, 59–77. Cambridge University Press: Cambridge University Press, 2013.

"An Ecomodernist Agenda," accessed January 18, 2016, http://www.ecomodernism.org/manifesto-english/.

Edgeworth, Matt, Dan deB Richter, Colin Waters, Peter Haff, Cath Neal, and Simon James Price. "Diachronous Beginnings of the Anthropocene: The Lower Bounding Surface of Anthropogenic Deposits." *The Anthropocene Review* 2 (2015): 33–58.

Ehret, Verna Marina. "Globalization, Redemption, and the Dialectics of Courage." *Sophia* 53 (2014): 67–80.

Heyward, Clare. "Situating and Abandoning Geoengineering: A Typology of Five Reponses to Dangerous Climate Change." *PS: Political Science and Politics* 46 (2013): 23–27.

Hulme, Mike. "Climate Change: No Eden, No Apocalypse." In *Exploring Climate Change through Science and in Society,* 204–6. New York: Earthscan/Routledge, 2013.

———. "Four Meanings of Climate Change." In *Future Ethics*, edited by Stefan Skrimshire, 37–58. New York: Continuum, 2010.

Jackson, Robert B., and James Salzman. "Pursuing Geoengineering for Atmospheric Restoration." *Issues in Science and Technology* 26 (Summer 2010): 67–76.

Jenkins, Willis. *The Future of Ethics: Sustainability, Social Justice, and Religious Creativity.* Washington: Georgetown University Press, 2013.

Klemm, David E., and William Schweiker. *Religion and the Human Future: An Essay on Theological Humanism.* Malden, MA: Wiley-Blackwell, 2008.

Lewis, Simon L., and Mark A. Maslin. "Defining the Anthropocene." *Nature* 519, no. 7542 (March 12, 2015): 171–80.

Lovelock, James. "A Geophysicist's Thoughts on Geoengineering." *Philosophical Transactions of the Royal Society* A 366 (2008): 3888–90.

Nerlich, Brigitte, and Rusi Jaspal. "Metaphors We Die By? Geoengineering, Metaphors, and the Argument from Catastrophe." *Metaphor and Symbol* 27 (2012): 131–47.

Olson, Robert L. "Soft Geoengineering: A Gentler Approach to Addressing Climate Change." *Environment* 54, no 5 (September/October 2012): 29–39.

O'Neill, Saffron J., and Nicholas Smith. "Climate Change and Visual Imagery." *WIREs Climate Change* 5 (2014): 73–87.

Petersen, A. "The Emergence of the Geoengineering Debate within the IPCC." Case Study, Geoengineering Our Climate Working Paper and Opinion Article Series. Available at:http://wp.me/p2zsRk-bp. Accessed January 18, 2016.

Preston, Christopher. "Re-Thinking the Unthinkable: Environmental Ethics and the Presumptive Argument against Geoengineering." *Environmental Values* 20 (2011): 457–79.

Ricoeur, Paul. "Narrative Time." *Critical Inquiry* 7 (1980): 169–90.

———. *Time and Narrative*. Three volumes. Chicago: University of Chicago Press, 1984–1988.

Scott, Peter M. "Humanity." In *Systematic Theology and Climate Change*, edited by Michael Northcott and Peter M. Scott, 108–23. New York: Routledge, 2014.

Skrimshire, Stefan. "Eschatology." In *Systematic Theology and Climate Change*, edited by Michael Northcott and Peter M. Scott, 157–74. New York: Routledge, 2014.

Smith, Bruce D., and Melinda A. Zeder. "The Onset of the Anthropocene." *Anthropocene* 4 (2013): 8–13.

Spence, Alexa, and Nick Pidgeon. "Framing and Communicating Climate Change: The Effects of Distance and Outcome Frame Manipulations." *Global Environmental Change* 20 (2010): 656–67.

Steffen, Will, Paul J. Crutzen, and John R. McNeill. "The Anthropocene: Are Humans Now Overwhelming the Great Forces of Nature?" *Ambio* 36, no. 8 (December 2007): 614–621.

Tillich, Paul. *Systematic Theology*, Volume 1. Chicago: University of Chicag Press, 1951.

Tillich, Paul. *Systematic Theology*, Volume 3. Chicago: University of Chicago Press, 1963.

Chapter Six

Rewriting Mortality: A Theological Critique of Geoengineering and De-Extinction

Stefan Skrimshire

Geoengineering has been described by some ethicists as a technological and geopolitical intervention necessitated by the failure of other policy measures (i.e., radical reduction of global carbon emissions) to curb global warming.[1] Thus, despite its explicit optimism about human ingenuity, the very appearance of the term has inspired such resignation as, to cite the forum held by the New America Foundation in 2010, the "Horrifying Idea Whose Time Has Come"[2] ; or the demand that we "think the unthinkable."[3] For some, even to consider engineering the climate marks a failure. Operating under the sign of failure of course raises its own subset of moral and political questions, such as who gets to decide the point at which failure has occurred, and what would count as a sufficient evidence for making this distinction? Making the case for geoengineering based upon the idea that we have run out of time for alternative solutions requires confident projections about complex political and social trajectories.

A popular criticism of geoengineering, championed by Naomi Klein, attempts to undermine this very assumption. Her claim is that what really drives geoengineering advocacy—referring specifically to solar radiation management (SRM) projects—is the idea that a technological solution offers us a sort of pacifying insurance against the disruption of the current economic and political order that truly exhausting "plan A" options (i.e., radical infrastructural change) would require. Geoengineering, on this view, holds off the end of the (capitalist) world: the future environmental risks associated with SRM are favorable in comparison with the *truly* unthinkable costs of confronting the end of our current levels of energy consumption and growth.

On the back of this novel ethical situation, we can sense already a new set of theological questions emerging. How are beliefs about human agency and the future of creation affected by this constructed narrative of failure? A suspicion underlying arguments such as Klein's is that geoengineering advocacy reflects a desire to literally capitalize upon the model of agency lamented in Bill McKibben's famous "end of nature" thesis and advanced in the light of the Anthropocene debate—namely, that human intervention is so entangled in the balance of Earth systems that we ought to accept our "post-natural" condition. We should "embrace our monsters," to cite Bruno Latour, and admit that our notion of failure must henceforth be reflexively managed[4] rather than absolutized. The parameters of any future "world" that we wish to preserve are irreversibly humanized. Forrest Clingerman puts it thus: "Geoengineering is argued as a form of transforming our environmental failures… not by redefining the human self in terms of being 'natural,' but by rendering the climate a domesticated participant, as it were, in such human betweenness."[5] Klein's argument against SRM technology would seem to suggest that not only the climate, but also and more specifically the world's poorest people become caught up in this experiment of domestication. This is because climate risks associated with the long-term effects of dimming the sun's rays will not be equally distributed. They will include disruption to summer monsoon periods and vastly extended periods of drought in sub-Saharan Africa. Klein's criticism is that geoengineering is premised on accepting some losses as necessary (disruption to weather patterns in developing countries) and others as nonnegotiable (curtailing civil liberties, imposing limits to energy consumption in the Global North).

In what follows, I want to offer a critical theological and philosophical perspective on the basic suspicion that Klein's argument touches upon. The suspicion is this: a consensus on geoengineering as a necessary response to a temporal emergency is premised not purely upon scientific arguments, but upon a cultural resistance to loss, or more specifically a resistance to the possibility of a meaningful ethical encounter with finitude and loss. To get there, I need to widen the context under consideration to include other technological responses to environmental challenges that invite us equally to rethink questions of human agency, finitude, and temporality. Geoengineering shares with other technologies an investment in temporal metaphors (something also discussed by Forrest Clingerman in chapter 5 of this volume). These metaphors are mobilized toward the belief that technologies can buy time, delay time (of runaway warming) and, in the case I want to consider, reach back into time as a way to confront ecological limits and correct our failures. Matthew Kearnes's climate reading of Foucauldian biopolitics—where "environmental politics is recast in biopolitical terms as a matter of rational administration, of literally managing a 'Future Earth'"[6] —parallels exactly the space of analysis I require to connect geoengineering with this wider consideration of the management of temporality in public discourse. For the narration of geoengineering as the techno-*management* of failure is related in important ways to scientific and technological responses to extinctions, the main consideration of this chapter. The technological promise of de-extinction projects—genetically "reviving" once extinct animal species, cloned using DNA samples from preserved cadavers—thus becomes immediately relevant to the discussion sparked by geoengineering. They both are implicated in the deliberations over the nature and acceptance of loss, which are also those raised in relation to geoengineering.

I want to develop a deeper understanding of the rhetorical and theological framework by which related discussions about naturalness and unnaturalness, timely and untimely losses, might be operating. One avenue for doing so involves looking at another theological metaphor in which natural plenitude is framed: the book of nature. It is commonly supposed that this metaphor, once popular to medieval theology as representing the "second book" of divine authorship, became obsolete in the modern period[7] in favor of mechanistic metaphors of nature. The examples of engineering I describe below challenge such a view. They suggest that the story of nature in the Anthropocene is particularly conceived as text, and raises once again the specter of Derrida's challenge—issued in *Of Grammatology*—to explore a form of writing *after* the "end of the Book"[8] in Western metaphysics. In the context that I want to consider, the book of nature has been not so much reopened as eternally rescripted and rendered readable and corrigible by successive forms of *techne* at the heart of a modernist reading of nature. We need to learn to read the end of the book alongside the end of nature. Because certain engineering interventions into environmental loss are narrated in terms that implicate the human (one might ask at this stage, human as author? editor? redactor?), specific questions arise about why and how irreversible loss—that is, extinction—might still be described as a form of writing.

If the Anthropocene represents a time of "extraordinary hypertrophy of hubris concerning the possibilities of predicting and controlling natural processes,"[9] we need to critique the sources of that mastery and their theological origins. The theological interlocutor I choose is particularly well versed in the ways we associate "the book" and material finitude. By introducing Saint Augustine and his interpreters into the debate, a consideration of textuality will be intricately linked to beliefs about transience.[10] The result will be to reflect on the mournful aspects of this metaphorical reading and writing and to see this aspect as driving an alternative to the techno-managerial and biopolitical response to losses wrought by climate change.

DE-EXTINCTION

De-extinction is the attempt by synthetic biologists to restore once extinct species through a variety of methods including cloning, genetic construction or "editing," and selective breeding. Real attempts have been made since 2008 to clone animals genetically proximate to, for instance, the extinct European Auroch.[11] Meanwhile, scientists from the Revive & Restore project in California are developing genomic technology that they believe will allow them to revive the iconic Passenger Pigeon, which has been extinct for one hundred years. Using skin samples from preserved museum specimens, scientists at The Great Passenger Pigeon Comeback project have been working to read the genetic data of the extinct bird in order to compare it to that of its closest living relative, the Band Tailed Pigeon. The surrogate pigeon will then be genetically engineered using this information to reconstruct a close resemblance to the extinct bird through "genome editing and germ line transfer." Other projects with much the same ambition, but attracting even greater publicity for their proximity to Jurassic Park fantasies, include the attempt to revive the Woolly Mammoth from frozen DNA samples discovered in the Siberian Tundra.

There is some rather obvious quasi-theological rhetoric foregrounding this work. Projects have been linked under the term "resurrection ecology" and use labels such as the Lazarus Project and the Ark Project. But the ethical dimensions of de-extinction raise further interesting concerns for theology because they rely upon a prior perception about what, in scientific terms, we call natural plenitude.[12] The loss of species extinction normatively considered—as a distinctive moral harm—relates to what we think we know about the totality of life "out there." The ethics of anthropogenic extinction thus involves deciding what counts as a normal or abnormal rate of loss—whether of habitats, species, individuals, or ecosystems—in order to prioritize remedial action. Which species, if any, *ought* we to bring back? Objections[13] focus on the biological and ecological viability of the once restored species (will it behave in the same way as its ancestors in the present habitat? What unintended impacts will its sudden existence have on the present ecosystems that have developed without it?), and the ethics of resource management (why not concentrate on protecting currently existing, endangered species?). The focus here is the management and control of loss.

Thus, in the same way that Klein argues that geoengineering's global risk assessment makes some lives dispensable, the selectivity of de-extinction prioritizes some lives over others. Its principal spokesperson, Stewart Brand, expresses this anthropocentrism overtly: "some species that we killed off totally, we might consider bringing back to a world that misses them."[14] What about those that the world doesn't miss? Klein's humanitarian worry haunts this parallel question. De-extinction from the outset targeted the possibility of resurrecting species that were not merely culturally iconic (the other popular example is the Thylacine, or Tasmanian Tiger, a figure of popular mythology and symbol of collective guilt). The chosen projects are themselves cultural reference points that reinforce, because offering to redeem, a narrative of human failure and mismanagement. For example, the extinction of the Passenger Pigeon is an iconic reminder of humanity's destructive capacity for overkill. The retelling of its story marks a major shift in human perception of natural plenitude. Once thought to be the most "abundant bird species in North America, if not the world" and which could "darken our skies for three days straight,"[15]a population of up to 5 billion was extinguished by overhunting and habitat loss, within a few decades. According to the popular narrative, the last pigeon, Martha, died in a Cincinnati zoo on September 1, 1914, where she is still on display.

An intriguing relationship between de-extinction and geoengineering emerges from this picture. The science and philosophy behind de-extinction projects are closely linked with re-wilding exercises: in particular, with attempts to recreate entire ecosystems that disappeared under the pressure of human impact and climatic change. The model of this idea is Pleistocene Park, a nature reserve in northeastern Siberia. In 1988, experiments by Russian scientists began to reintroduce flora and fauna that re-create over time the ecosystem of the "Mammoth Steppe." Most significant of the project's scientific premises, for my purposes at least, is the belief that the Mammoth, Woolly Rhino, and other megafauna that once roamed this region were killed off by early humans (as opposed to the competing theory that this was initiated by climate change). It is believed that the ecosystem that supported them played a key role in mitigating against global warming. The argument is that grazers like the Mammoth, Musk-ox, and Yakutian wild horses maintained the grasslands of the Steppe, and that these grasslands were some of the most efficient recyclers of water and chemical nutrients, sequestering a great deal of carbon dioxide. Sergey Zimov, the cofounder of the project, has claimed that the former Pleistocene plains hold more carbon in their soils than all of the world's rainforests.[16] Engineering the Mammoth's ecosystem has thus become tied to a strategy for mitigating against catastrophic climate change by preventing the release of these stores into the atmosphere. Re-wilding and de-extinction, both rightly interpreted as feats of engineering in their own rights, are thus interrelated. The Pleistocene ecosystem restoration continues by gradually integrating more wild herbivores, and even via synthetic re-creation of megafauna activity by driving ex-Soviet tanks across the landscape to mimic the impacts of Mammoths upon the vegetation.[17] This vision of Edenic harmony to which humanity is called to return goes beyond traditional visions of preindustrial or even pre-Columbus humanity. It stretches to the very arrival of late Pleistocene/early Holocene *Homo Sapiens,* around twelve thousand years ago.

On these grounds at least, it has been argued that de-extinction and rewilding projects, taken together, constitute a type of geoengineering. They would not, perhaps, meet the criteria demanded by some, of a "deliberate large-scale manipulation of the planetary environment."[18] Nevertheless, they both represent a claim about the role of human agency in response to a narrative of mortality and loss. For geoengineering and de-extinction both presume that technological ingenuity presents humanity with the obvious opportunity to account for its moral failings elsewhere. It is no surprise, therefore, that Stewart Brand has been an influential public advocate of both. Brand is best known as an early pioneer of pragmatist environmentalism, combining faith in scientific innovation with liberal politics. His *Whole Earth Catalog,* an open resource first published in 1968, is still hailed in the United States as the bible of DIY 1970s self-sufficiency. In its latest incarnation three decades later, the same pragmatist spirit has been applied to the management of the earth system itself: *The Whole Earth Discipline: Why Dense Cities, Nuclear Power, Transgenic Crops, Restored Wildlands, Radical Science, and Geoengineering Are Necessary.*[19] As with the salvationist rhetoric of de-extinction, Brand's defense of geoengineering is also famously narrated, in the words of the first *Catalog*, through the divinization of the humanity's agency: "We are as gods and have to get good at it."

Klein makes much of the fact that this brand of environmentalism emerged alongside the first picture of planet earth viewed objectively (since 1969) from space. What had been viewed as a prerequisite for an ethic of care and stewardship became, in Klein's view, replaced with the engineer's ability to abstract the earth as a toy to manipulate, break, and mend as the situation dictates.[20] It is also no coincidence for Klein that Brand was also one of the early advocates, since the 1970s, of colonizing Mars. This idea was again inspired by the newfound ability to view earth as an abstracted entity, and one that we might one day leave behind. But what really connects the rhetoric of each of Brand's pet projects—Mars colonization, geoengineering, and de-extinction—is their reliance upon humanity's corrective agency over the destiny of creation. Resurrection ecologies, in the assessment of the historian of science Stephanie Turner, "strongly dramatiz(e) the mythos of human beings standing in for the divine in caring for nature, including the salvific work of recreating it."[21] A picture of natural plenitude is often evoked—explicitly, in the case of the artwork that adorns the pages of Revive & Restore (see figure 6.1)—that assumes a (contested) myth of a prehistoric harmonious wilderness. Brand's rhetoric confirms this: "Humans have made a huge hole in nature over the last 10,000 years. We have the ability now and maybe the moral obligation to repair some of the damage."[22]

Environmental biopolitical management can be illuminated through political theology as, in Kearnes's words, the "interplay between the administrative and the miraculous."[23] If this is so, then both geoengineering and de-

Figure 6.1. Isabella Kirkland, "Gone," oil paint on canvas.

extinction reveal a particular temporal focus toward the miracle. That is, an intervention into the normal time frames of ethical and political deliberation. A further aspect of Brand's illustrious career has been his cofounding of the Long Now Foundation, the project which funds many of the de-extinction initiatives in the United States, including Project Passenger Pigeon. The foundation's remit is to "get people thinking past the mental barrier of an ever-shortening future"[24] through projects that enable long-term thinking, predicting, and engineering. On top of geoengineering research and de-extinction projects, these include the art installation Clock of the Long Now. Funded by Amazon's founder Jeff Bezos, the clock is currently under construction and destined for an underground location in the Nevada desert. It is designed to continue ticking, once every year, for ten thousand years. The moral significance of "long-term thinking" for Brand and associated writers is that it provokes moral considerations for a far future that we won't inhabit. Brand thus believes that the image of the Clock of the Long Now will do for thinking about time what the image of the whole earth did for thinking about nature.

But such an argument is precisely, and starkly, absent from the advocacy of its sister projects, geoengineering and de-extinction. One of the most convincing moral objections to solar dimming proposals has been in response to its reliance upon continued human maintenance. In the case of injecting sulphur particles into the atmosphere to deflect solar radiation, any sudden interruption of this injection would precipitate a sudden and catastrophic warming. Thus, as a technological response to continued human planetary interference, it requires certainty about favorable human governance. A similar point can be made in relation to de-extinction and rewilding exercises. Such management of resurrected and prolonged ecosystems will, if the current evidence of Pleistocene Park is an indicator, rely upon continued monitoring and adaptation of techniques based upon our human control and management of those systems. Both projects and their appeals to the miraculous are premised upon the continued projection of human technical management—the prolongation of a contemporary sense of human mastery.

REVIVING THE DEGRADING BOOK OF NATURE

At this point I wish to explore some of the possible theological sources of critique in response to the metaphors and narratives sketched above. De-extinction, as we have seen, is premised on a narrative of human weakness that grants the function of modern science a uniquely redemptive agency. In theological terms this narrative may reflect a kind of materialist postmillennial eschatology. In such a vision, humanity has a special hand in establishing and witnessing the coming of God's kingdom by shaping both human and natural history. There are strong precedents for charting the legacy of postmillennialist belief in developments of scientific intervention into the secret knowledge of creation, from early modern science onwards. Francis Bacon famously framed the revelatory, apocalyptic task of the new science as decrypting the book of Nature[25] in ways strongly reminiscent of contemporary de-extinction advocacy. To cite one example, the title of a flagship publication advocating de-extinction by biologist George Church—*Regenesis: How Synthetic Biology Will Reinvent Nature and Ourselves*—points toward a return to the original text and act of creation. The same work refers to the science of genome sequencing in one chapter heading as the art of "Reading the Most Ancient Texts."[26]

Stephanie Turner, charting the history of scientific extinction narratives, observes that such discoveries have often been thus described, combining the metaphor of text with an eschatological temporal model accessible to scientific knowledge. By redefining the natural boundaries of the lifespans of created beings, human history is implied not only in natural history but also in the very ends that are prefigured by the plenitude of creation. Extinction is in some significant sense a terminus, of course. But an "infinite mutability" has crept in because DNA legibility now provides geneticists the chance to imagine that extinct species may live on. This is not only because of the possibility that their species might be one day brought to life. It is also because when their DNA are read as permanent, preserved features of the world, the life of dying things is inculcated into a frozen, "perpetual present. . . . The endless reproducibility of DNA renders all extinction open-ended. Both conceptually and informatically (that is, in terms of the posthumous existence of genetic code in a computer database), extinction is becoming less and less absolute."[27] The readability of genetic sequences, even if only hypothetically viable, represents a significant challenge to the visibility of death, and also its finality.

Turner calls this "historical revisionism." It indicates a shift in what counts as natural death and in what we are to think, ethically and scientifically, about species disappearance. On the Revive & Restore website, under a dedicated page entitled Extinction Continuum, a pictorial timeline encourages us to visualize this as a spectrum. (See figure 6.2.) It begins with the present age of "endangered" species, through to shadow images of relatively recently extinct species who's DNA is recoverable. Next in line are those whose viability for rescue is fast disappearing because of "degrading DNA." Finally, we see the dinosaur skeleton bearing the suddenly darkened label, "No DNA." The definitive baseline is no longer that which separates existence and disappearance. It is that between "viable" and "degraded" DNA.

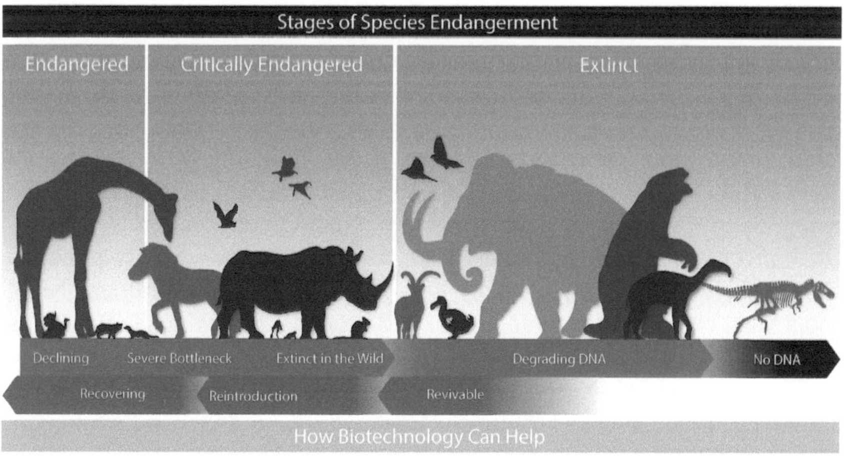

Figure 6.2.

"No DNA" means "no future." But up until this point, the animals from the past appear to us in vibrant, animated imprints, explicitly *not* as fossils. They are in an important sense "alive" for us. Most importantly, they are alive in their accessibility as text, in their readability. Jay Clayton[28] reminds us that DNA is essentially a "sign system" in which four letters enable the entirety of biological variations to be sequenced and assembled, including genetic traits that tell a story about the past and future of that organism. This affords gene science the privileged position of presenting nature as an open book. Certainly the engineers behind the Great Passenger Pigeon Comeback are aware of the potency of this textual metaphor, and capitalize upon it.

The technique currently being pursued for the passenger pigeon revival is known as "in vitro primordial germ cell engineering." This has been described in simplified terms as making "germ edits" in petri dishes, much like a writer at his desk. Assembling the genome of the surrogate species to be compared is consequently called a "first draft."[29] But rather than simply evoking the role of sequencer as corrective editor of the past, de-extinction engineers are also clear in their belief that the technology represents the opportunity to construct a better future. As Ben Novak, the geneticist working at Revive & Restore, puts it:

> Contrary to the poetic nature of "righting past wrongs" that some attribute to the de-extinction of passenger pigeons, I view [it] as a project seeded in our present and future; it is a pivotal exercise in thought stressing the recognition that we are the drivers of change on this planet and that we have the cognitive ability to take responsibility for the direction of that change. . . . We ultimately write the story. Martha's story is a call to action still, a century after her death. Her story is far from over.[30]

ECO-THEOLOGY OF BODIES AND BOOKS

Obviously, genetic rescue does not involve the resurrection of an individual creature. Martha the pigeon—*that* particular pigeon—is dead, and will not return to life. Moreover, to date, certainly the cloning technology has not promised the resurrection of exact copies of long extinct creatures, only the reintroduction of their specific traits into existing species.[31] Nevertheless, what advocates of this technology hope to revive is a genetically proximal identity once thought irrecoverably lost. So, de-extinction overturns a belief about what counts as a natural process and grounds upon which to mourn its passing. This reinvigorates questions that are central to Christian theology. In particular, a central concern of eco-theology has been critiquing traditional narratives of humanity as "co-creator" and guardian of the world's future. A prominent feature of feminist contributions to such discourse since the 1980s has been an exposition of the patriarchal and imperialist framing of Christian eschatology. Feminist theologians have posed questions such as these: to what extent are orthodox doctrinal visions of the future new heaven and new earth bound to project our fantasies of the survival, uncorrupted, of "our world" into eternity? Is the belief in immortality a typically male fantasy? And to what extent is it within the theologian's remit to reconnect social practices with the opposite of this faith—with the realities of transicncc, and the irreversibility of death in this life?

Rosemary Radford Ruether's *Sexism and God-Talk* is the key example of this latter suspicion. Ruether links much of the classic patristic treatises on eternal life and the resurrection of the body with a certain discourse about male bodies in particular. She makes a further association between this imperialist fantasy of immortal victory of the human over corruptible nature and the transience of natural history. In this analysis, female nature as grounded in cycles of death and rebirth historically became associated with sin. Natural biological cycles, menstruation and childbirth, carried with them the "taint of decay and death"[32] in the Christian imaginary.[33] It is well known that prominent second-century Christian apologists argued that the resurrection of the body—and thus its removal from that world of decay—represented the completion of human potential.[34] Many specified that women would be reconnected with male bodies on judgment day, since their own bodies were imperfect versions of this apex of creation.[35]

These reconsiderations of classic sources of theological debate can be brought to bear on the contemporary dilemma I have sketched above. What is the relationship between this materially fleshy hope in the restoration of creation (a rematerialized, plenitudinous Ark) and the eschatological debates about the resurrection of the body? Is there something of the early Christian's belief in the resurrection of the body, and the veneration of material flesh (of Christ in the Eucharist, and holy martyrs in relics) in the redemptive offering of the dead flesh of extinct species? A key feature of de-extinction projects is its mythic narrativization: a heroic quest to retrieve the very origins and stuff of life that offer redemption to all species. The romanticized story of the hunt for the Mammoth (for the purposes of DNA cloning) by Russian scientists across icy tundra is an example of this very search for the holy grail of preserved flesh that bears the code to new life.[36] The powerfully evocative image of Martha the Passenger Pigeon is another example.

But such theological debate can also penetrate further into the cultural metaphors we are witnessing here in relation to the link between salvific flesh, humanity taking on the mantle of mortal authorship, and the metaphor of the book. Andrea Nightingale has argued that the materiality of books—from reflections upon the eternal law of holy scripture, to the popular traditions of hagiographic literature—are central to St. Augustine's *Confessions;* and that the book's theological significance still resonates with contemporary social and moral debates about mortality and finitude. Through Augustine we

Figure 6.3. Preserved skins of passenger pigeons (Ectopstes migratorious) in a specimen drawer at the Museum of Natural History in New York.

have access to the very first experiment in autobiography in Western thought. The confession itself is an exercise in questioning our desire to record "the truth" of the book (why confess to God, he wonders, when God already knows everything?) alongside the necessarily material and corruptible nature of words. Continental philosophers from Derrida to Ricoeur have been quick to point out the "proto-postmodern" implications of such a reflection.[37] Books, on this Augustinian view, are one more site of interaction between the communication of unchanging truth (the eternal Word) and the finite contextuality of its creation. In Nightingale's synthesis, there is a crucial link between this attention to books and the role of bodies more generally, what she calls the interchangeable functions of corpse and corpus. This is because in late antiquity it is the flesh of human bodies (of saints and martyrs, holy men and women) that take on a kind of mediating role between the materiality of the body and its heavenly destination.[38] The Bible is the ultimate paradox of the word: a transient object that yet imparts the eternal, unchanging truth of God. In an extraordinary metaphorical description of this mediating role, Augustine imagines God stretching the "skin" of his Book as a firmament over the world in order to nourish creation with his Word. Creation itself is thus understood as clothed in and *as* a book of life.

The most telling association of text and creation is contained in Augustine's phenomenological discussion of time in chapter 11 of the *Confessions*. Here he explores the nature of time through the experience of language. Like the recital of words, we can measure time only through its passing from present into past—and "recorded" in us only via memory. This mournful aspect to language and memory is directly related to our fallen condition. Memory itself, being distended in different directions and illusive in its presentation of facts of our life, is a frustrating condition. It is the result of our fall from the perfect condition of heavenly bodies. Words themselves, discourse bound by material, physical limits of dissemination, comprehension, and retention, thus bear the mark of sin and finitude. Even biblical truth can only be received with difficulty through, and in, the deaths of generations of prophets.[39] This is a legacy that leaves open different, temporally mediated versions of the truth. More profoundly still, according to Augustine, the mortal destiny of all created things, collected only in the memory of the creator, are related to this transient act of speech. Through speech words come and go, but give meaning only to the one who hears the whole sentence.

The powerful associations that emerge for our discussion about confronting textuality and finitude recall the following passage from the *Confessions*:

> Not all reach old age, but all alike must die. When they set upon the course of their existence, and the faster they climb towards its zenith, the more they hasten towards the point where they exist no more. This is the law they obey. This is all that you have appointed for them, because they are parts of the whole. Not all the parts exist at once, but some must come as others go, and in this way together they make up the whole of which they are the parts. Our speech follows the same rule, using sounds to signify a meaning. For a sentence is not complete unless each word, once its syllables have been pronounced, gives way to make room for the next. (*Confessions* 4.10.15)

Here the metaphor of language is employed to describe a theology of transience. It is our unavoidably temporal engagement with the material world that makes its reading a mournful activity.

With this in mind, we can now return to some of the questions haunting eschatologically oriented eco-theology. Ernst Conradie has attempted to articulate Christian belief in the resurrection, cognizant of the feminist critique with which I began—that is, cautious of the immortalist fantasies of preservation and survival "beyond" natural cycles of death and decay. At the same time, he claims that the eco-theologian wants to retain from early Christian tradition the belief that something of the goodness of corporeal existence is preserved in eternity. As St. Paul had put it, through the resurrection the "perishable becomes imperishable" (1 Cor. 15:53).[40] We should, indeed, want to resist any account of an eschatological future that absolves the present of its obligations to creation, including the preservation of endangered species. In a move oddly reminiscent of Augustine, Conradie's solution is to suppose that the eschaton, operating at a different cosmic dimension to human time, is best described through the analogy of "material inscription." He asks us to imagine that the whole history of the cosmos is inscribed materially, in the manner of a "book of life" with its material particularity, upon the mind of God. In the eschaton, in which all things are finally redeemed, the book provides the "building blocks" that are preserved in God's memory. The result is that the Christian might valorize the significance of their acts in the mind of God without "nullify[ing] human mortality and transience" because each finite moment is preserved for eternity.[41] The significance of our lives—and by extension those of all species in the world—are contained in some act of divine memorialization. This has its moral flip side, of course. Conradie confronts the implication that all moments of death and pain are thereby preserved. All is remembered, and judged, presenting as much a vision of hell as of hope, and requiring the supplement of Christ's salvation.[42]

What is going on in the employment of this metaphor? Such a move, it is hoped, retains the Jewish apocalyptic element of Christian eschatology—that is, a revolutionary hope in the incoming future as the transformation of material history (in line, I would suggest, with Jürgen Moltmann's ecological eschatology). At the same time, it refuses the erasure of memory that would be implied in the wiping away of all materially embodied past wrongs. A further metaphor is used of God retrieving the "photographs" that exist of our finite experiences for eternity in the book of life, preserving the happy and tragic moments alike. God nostalgically flicks through the photo album, witnessing the ups and downs of creation. He is able to judge, lament, and rejoice in equal measure. Conradie's eschatological vision is meant to reconcile Christian hope with a politics of memory and justice. His wish is to preserve eschatology as a resource for ethical thinking in the present.

Yet I would argue that this photographic metaphor equally invites an association that represents a reverse of this historicized eschatological hope. Walter Benjamin knew very well the melancholic structure of this photographic encounter with the world and history. Like the early Daguerreotype photographic portraits that attempted to preserve the melancholic allure of its subjects as if frozen in eternity as subjects of our gaze, there is something "corpse-like" in the frozen image of the photo and our obsession with preserving it. More generally, we ought to be wary of what this return to the metaphor of the readable book of natural history says about our engagement with the Anthropocene. Would the believer—in adopting Conradie's framework—be seeking an eternal accountability (with the creator) through photographic memory? Or would they be fantasizing their own immortality through the readability of the earth and their actions upon it?

FOR A THEOLOGY OF TRANSIENCE

What might this theological discussion about materiality, finitude and textuality offer to a debate about de-extinction and geoengineering? The critique of our human attempts to access the pure "book of nature" in pursuit of material perfectibility invites a typically Augustinian insight: we are not God, and thus we are not master readers of the book of nature. Rewriting mortality relies upon this sense in which humans have access (through DNA decoding) to some original text or, perhaps, some purer text newly decoded.[43] A crucial insight that therefore has been lost from the theological nuances of the metaphor is the book as transient and perishable entity rather than as signature of the immortal archive. Augustine would certainly warn us against venerating the flesh of a pigeon, nevertheless. More problematically, he might also reprimand us, as he did himself in the *Confessions*, for enjoying the life of any living birds (or any animal) at an aesthetic level. Birds are part of a transient, perishing world. He has room for veneration *out of* the encounter with impermanence, however: "Let these transient things be the ground on which my soul praises you (Ps. 145:2), "God creator of all." But let it not become stuck in them and glued to them with love through the physical senses" (IV; 15). So the attachment to transience becomes both the necessary prism through which our love (for God) is known and experienced, and a warning against confusing our "true love" with a desire for permanence.

The tension between being earth-bound and also "out of nature" finds ultimate expression in Augustine's acts of mourning the death of loved ones. In the *Confessions*, Augustine's visceral grief at the death of his friend plays an essential part in his illustrating the split nature of man's temporal condition. Witnessing the death of that which we love rends us in two. One part of us is bound to things that are passing, the other hopes for a future redemption. Grief is this condition of mental conflict: "I had poured out my soul on the sand by loving a person as sure to die as if he would never die" (IV;13). Augustine here seems to join the ranks of those who would admonish us for grieving. As Nightingale notes, whereas Augustine's lament for his friend is performed in "fear and loathing," finding no relief, his grief over his mother, still fierce, finds reconciliation that her death is not meaningless but has a function. Augustine's grieving actions become literally more measured, evenly aggressively so: rebuking his son for crying at her deathbed, stopping the welling of his own tears by "violent command of my will," and later allowing himself to cry for a "fraction of an hour" (*Confessions* IX: xii).

Confusingly, this would seem akin, invoking Freudian terms, to an instance of successful mourning in the clinical sense: letting go of the libidinal attachment to the lost object. Conversely, one would have grounds for associating Augustine's *initial* reaction to loss with that of a melancholic, or "failed" mourner, since in both cases it is the fixation with the lost object—its refusal to be overcome—that is identity forming.[44] However, it can be argued that Augustine is not simply exercising a macho control over his emotions. He is, rather, highlighting misplaced attention. The true object of our love (God, and eternity) is lost in our own midst. It cannot be found *here* in the temporal sphere by replacing one transient object with another. Thomas Carlson captures this in his analysis of Augustine's critique of the love of "world": "The human temporality that imitates God's eternity in this way, like the sin from which it derives, wherein we imitate perversely the freedom and independence of God, goes hand in hand with a false security, or self-satisfaction. . . . It is from this perspective that Augustine will warn against the desire for security in this world and insist that a trembling heart in the face of our peril should replace all self-complacency."[45]

A powerfully melancholic reading of the situation I am considering presents itself. This recalls in particular the rhetoric of human power and potential over the finite limits imposed by humanity's own failings. This sense of power was overt in both Brand and Novak's rhetoric, cited previously. A sense of lostness is also implied in this inability to regain ourselves wholly in this life. The aesthetic plea of Novak's defense against de-extinction's moral critics betrays something of this lack, an inability to contemplate loss: "A new passenger pigeon may cause an internal debate in the on-looker, but when the piercing red iris meets the on-looker's gaze, in a moment of deep connection, it will remove all doubt, as if to say, 'I am the passenger pigeon reborn'" (Novak 2014).

Some implications of this Augustinian reading can be drawn here, as much for geoengineering as for de-extinction (indeed, they show the relation between the two quite well). Suspicion of this desire for security is related to the critique of technological mastery over finitude, inasmuch as these would replace our obligations to the living in acts of justice and peace. For instance, a very credible fear recognized by both advocates and opponents of geoengineering is that even *discussing* the potential promise of implementing geoengineering solutions may act as a huge disincentive to pursue social and political measures to reduce carbon emissions.[46]

Stewart Brand's response to this worry has been to enlist the proposal for "patient geoengineering" from the physicist David Keith, a fierce defender of geoengineering:

> The common worry about geoengineering is that because it is so cheap ($1 billion a year) and easy, civilization would become addicted and have to

> continue it forever, while giving up on the expensive and difficult process of reducing greenhouse gas emissions, thus making the long-term problem far worse. Keith's solution is to design the geoengineering program as *temporary* from start to finish. 'Temporary' means shut it down by 2200.[47]

Aside from suspecting such plans of political naiveté, we can recognize in Brand's unflinching confidence in technology a basic mastery over time-scales (to be "temporary from start to finish" is a curious tautology. Who declares the start and end points of these technologies?).

We might also recognize in this confidence elements of what Heidegger targeted through his technological critique of modernity. That is, of the acceptance of technology as at once an act of revealing the world (as calculable, manipulable, readable) and at the same time an absolute masking of finitude and material limit.[48] The nihilistic technologization of life, inasmuch as it tends to conceal our basic finitude, thus removes *Dasein* from the existential potentialities that are opened up by being-towards-Death, its encounter with limit. Carlson's joining of Heidegger and Augustine might thus usefully be applied to my own critique of the technological management of ecological temporality: "[Augustine and Heidegger] both suggest, on the one hand, a love or care that really hates and gives death by closing any genuine future in the name of an actuality that we might produce, manage or consume within the economies of our self-possession (roughly, Augustine's *cupiditas*); and on the other hand, a love or care that loves authentically and gives life its chance by giving and keeping open a futurity that escapes our capacity for production and management and consumption."[49] This recourse to an ethics of care via a critique of death-defying technological mastery also recalls Ruether's critique of patriarchal eschatology. That critique was suspicious of any desire to extend the empire of material goods into particular Anthropocene vision of the eschatological future. The error to which "love of the world" makes people prone is that it reduces the horizon of human happiness and hope to that which the human, alone, can master, control and preserve.[50] This elevates the role of human technical capability as the guarantor of a peace that is sought for self, society, and world. A love that only reproduces sources of mastery, *libido dominandi,* substitutes the work of mourning for new forms of human governmentality.

CONCLUSION: THE EMBARRASSMENT OF LAZARUS

Brand's promotional de-extinction TED Lecture, with the iconic image of Martha the passenger pigeon in the background, begins with the tragedy of mass extinctions. It ends with the rallying cry commonly attributed to Joe Hill, a labor activist who was executed in the United States in 1915: "Don't grieve. Organize!" In Brand's use of it, the phrase indicates a macho admonishment against the act of grieving. Mourning is misplaced labor in the light of the promises of engineering.[51] Why weep when you have the power to resuscitate, to overcome the power of death?

This question haunts the underlying rationale to virtually all de-extinction projects. One of the most recent, which promises to revive and restore the extinct Australian Gastric-brooding Frog, is named the Lazarus Project. It is perhaps ironically fitting that, like the biblical Lazarus, the frog was technically resuscitated, only to die shortly thereafter of natural causes, none of the embryos having survived beyond a few days. The technological "success" of the experiment, nevertheless, was enough to earn its scientists the accolade of "one of 2013's best inventions."[52]

In *The Testament of Mary,* Colm Tóibín retells the story of Lazarus through the eyes of an all-too-human mother of Christ. Here, the tale is a scandalous affront to our sacred relationship with mortality: "No-one should tamper with the fullness that is death. Death needs time and silence. The dead must be left alone with their new gift or their new freedom from affliction."[53] Was Jesus's action in raising Lazarus really to show off the omnipotence of God against any law (not only natural law, but also the law of Sabbath observation, in the face of his orthodox Jewish audience)? John Caputo has argued that such a metaphysical leap of imagination undermines entirely the profounder element of John's story, that "weakness of God" contained in the famously shortest sentence: "Jesus wept." Tóibín's Mary says that Lazarus came back to "prepare to die" once more, and revealed in his way that this preparation is an ethical task linked to our confrontations with mortality.

We can consider this theological problem in light of de-extinction, particularly in view of the religious metaphors by which it is contested, and of the resistance it has provoked (by Novak, Brand, and others) in defense of a narrative of human technical mastery. One interpretation was that, through the example of Martha the pigeon, the museum specimen has been given a new redemptive role. The corpse becomes a relic or, in Freudian terms, a surrogate object. If we take Brand's activist rhetoric as indicative, the cultural legacy of this transformation is also a visceral resistance to the act of mourning, which is seen as a form of emasculation, of weakness in the face of a planetary emergency. But the ethical imperative in the work of mourning—by which our relationship to others is transformed—is also being offered and is articulated as dwelling with loss.[54] This is a topic of the Environmental Humanities that calls for greater theological engagement.

My own view, which I have tried to develop here, is that, in addition to its ethical critique, we need to understand powerful theological narratives that underwrite such a perspective and those that offer a possible critique. By looking at the metaphor of the book of nature in particular, we can better appreciate the way a concept of human agency in decoding and rewriting the text of life continues in the forms of earth systems engineering with which we are concerned. It is thus interesting to note in relation to this that rising political and moral concerns over geoengineering coincide with a revitalization of interest in deconstruction theory in relation to the Anthropocene.[55] The critique of technologization of nature involves once again unearthing and troubling the notion of original text, the true Word or "divine writing," to which we may return or even improve, through our acts of corrective editing.

In concluding, let me revisit the point raised at the start about geoengineering and de-extinction operating under the sign of failure. The critique of the capitalist-driven motivation for much of geoengineering and de-extinction advocacy is precisely—to take seriously Klein's argument—that these advocates are really resisting the death of the neoliberal world by stepping in where nature's limit ends. Bruno Latour made this point in reference to opposing uses of science to defend one's particular worldview in relation to climate change: "Climate deniers have clearly realized that it is indeed an end of *their* world. And, quite reasonably, they resist it."[56] A rejection of the act of mourning in Brand's example is, in the context of this insight, not only to be viewed as a macho response to death itself. It is also, and perhaps more precisely, the technocratic and managerial hubris that maintains the continuation of *our world* far into the future, and thus a resistance to declaring its death prematurely. What opponents of geoengineering are likely to share with those of de-extinction is a concern with what this resistance means for ethical commitment to the present. In both cases, our unwillingness to mourn environmental loss may reflect our incapacity to do justice in the world that we have.

NOTES

1. Stephen Gardiner, "Is 'Arming the Future' with Geo-engineering Really the Lesser Evil? Some Doubts about the Ethics of Intentionally Manipulating the Climate System," in *Climate Ethics: Essential Readings*, ed. Gardiner et al. (Oxford: Oxford University Press, 2010).
2. Naomi Klein, *This Changes Everything: Capitalism vs the Climate* (London: Allen Lane, 2014), 263.
3. Christopher Preston, "Re-thinking the Unthinkable: Environmental Ethics and the Presumptive Argument against Geoengineering," *Environmental Values* 20 (2011): 457–459. Matthew Kearnes has also reflected critically on this narrative framing of geoengineering as failure. Like others, Kearnes cites Paul Crutzen's influential essay defending research into atmospheric sulfur injections as a possible "escape route" in the event of unstoppable global warming, as providing the "legitimate" scientific narrative that kick started such a discourse of the emergency/last resort. Matthew Kearnes, "Miraculous Engineering and the Climate Emergency: Climate Modification as Divine Economy," in *Technofutures, Nature and the Sacred*, ed. Drummond, Bergmann, and Szerszynski (London: Ashgate, 2015).
4. Forrest Clingerman, "Geoengineering, Theology, and the Meaning of Being Human," Zygon 49 (2014).
5. Clingerman, "Geoengineering," 12.
6. Kearnes, "Miraculous Engineering."

7. Forrest Clingerman, "Reading the Book of Nature: A Hermeneutical Account of Nature for Philosophical Theology," *Worldviews* 13 (2009): 72–91.

8. I am indebted to the analysis that Bron Szerszynski (2012) makes between climate change, meteorology, and Derrida's *Of Grammatology*.

9. Bronislaw Szerszynski, "Reading and Writing the Weather: Climate Technics and the Moment of Responsibility," *Theory, Culture and Society* 27, 2–3 (2010): 22–23.

10. Forrest Clingerman also emphasizes (2009), along with the book as revelation, the material nature of textuality.

11. Jacob Sherkow and Henry Greely, "What If Extinction Is Not Forever?" *Science* 340, no. 6128 (April 5, 2013): 32–33.

12. I owe this insight and general discussion to correspondence with Callum Roberts, Professor of marine conservation biology at the University of York, UK.

13. See for instance Ron Sandler, "The Ethics of Reviving Long Extinct Species," *Conservation Biology* 28, no. 2 (2013): 354–360.

14. http://longnow.org/revive/.

15. http://passengerpigeon.org.

16. Peter Guest, "Giant Cows, Geoengineering and the Lazarus Frog," *Surviving the Future* https://medium.com/surviving-the-future. Accessed June 3, 2015.

17. Matthew Chrulew, "Reversing Extinction: Restoration and Resurrection in the Pleistocene Rewilding Projects," *Humanimalia: A Journal of Human/Animal Interface Studies* 2, no. 2 (2011): 19.

18. David Keith, quoted in Forrest Clingerman, "Redeeming the Climate," in *Technofutures, Nature and the Sacred,* ed. Drummond, Bergmann, and Szerszynski (London: Ashgate, 2015).

19. Stewart Brand, *The Whole Earth Discipline: Why Dense Cities, Nuclear Power, Transgenic Crops, Restored Wildlands, Radical Science, and Geoengineering Are Necessary* (London: Penguin, 2009).

20. Klein, *This Changes Everything*, 286.

21. Stephanie S. Turner, "Open-ended Stories: Extinction Narratives in Genome Time," *Literature and Medicine* 26, no. 1 (2007): 61.

22. http://longnow.org/revive/.

23. Kearnes, "Miraculous Engineering," 227.

24. Danny Hillis, quoted in Stewart Brand, "About Long Now," http://longnow.org/about/.

25. Stephen Matthews, "Francis Bacon's Scientific Apocalypse," in *The End That Does: Art, Science, and Millennial Accomplishment,* ed. Cathy Guttierez and Hillel Schwartz (London: Equinox, 2006), 97–115.

26. George Church, *Regenesis: Reinventing Nature and Ourselves* (New York: Basic Books, 2012), 37.

27. Stephanie Turner, "Open-ended Stories: Extinction Narratives in Genome Time," *Literature and Medicine* 26 (2007): 59.

28. Jay Clayton, "Genome Time," in *Time and the Literary,* ed. K. Newman, J. Clayton, and M. Hirsch (London: Routledge, 2002).

29. http://longnow.org/revive/.

30. Ben Novak, "Reflections on Martha's Centennial," 2014. http://longnow.org/revive/category/passenger-pigeon/.

31. Beth Shapiro, *How to Clone a Woolly Mammoth: The Science of De-extinction* (Oxford: Princeton University Press, 2015), ix.

32. Rosemary Radford Ruether, *Sexism and God-Talk* (London: SCM Press, 1983), 248; *Introducing Redemption in Christian Feminism* (Sheffield: Sheffield Academic Press, 1998), 116.

33. For a discussion of systematic theological responses to Ruether, in particular with reference to Moltmann, see Ernst Conradie, "Resurrection, Finitude, and Ecology," in *Resurrection: Theological and Scientific Assessments,* ed. T. Peters, R. J. Russell, and M. Welker (Grand Rapids: William B. Eerdman's, 2002). Christian eschatology, drawing upon ideas from Augustine but developed by the likes of Pannenberg and Moltmann, attempts a kind of bulwark against Gnostic pessimism about the future of material creation. It is the hope in the future life of human flesh that makes sensible, for instance, to a thinker like Irenaeus, the focal perfor-

mance of Eucharist as the faith in the real incarnation of God (ibid.). If God saw that it was good, why would he abandon creation through the irreversible processes of natural decay?

34. Brian Daley, "A Hope for Worms: Early Christian Hope," in *Resurrection: Theological and Scientific Assessments,* ed. T. Peters, R. J. Russell, and M. Welker (Cambridge: William B. Eerdmans, 2002), 140.

35. Andrea Nightingale, *Once Out of Nature: Augustine on Time and the Body* (Chicago: University of Chicago Press, 2011), 46.

36. Matthew Chrulew, "Hunting the Mammoth, Pleistocene to Postmodern," *Journal for Critical Animal Studies* 9 (2011).

37. John D. Caputo and Michael J. Scanlon, *Augustine and Postmodernism* (Bloomington, IN: Indiana University Press, 2005).

38. According to Nightingale, against a suspicion that, once "out of nature," postfall, the human looks only toward the death and overcoming of physical life, the mature Augustine offered a theology of the body that came to emphasize wholeness and orderliness of body and soul. The ascetic, mortified, and martyred body presents an intriguing exception to the rule that dead flesh reflects the fallenness of creation. This is because the miraculous status of the martyr's body brings the believer closer to God and eternal life, not away from it.

39. Nightingale, *Once Out of Nature*, 136.

40. Conradie, "Resurrection, Finitude, and Ecology," 291.

41. Ibid., 292.

42. Ibid., 293.

43. For the link between Derrida's *Of Grammatology* and St. Augustine's *Confessions*, see Mark Vessey, "Reading Like Angels: Derrida and Augustine on the Book (for a History of Literature)," in *Augustine and Postmodernism,* ed. J. Caputo and M. J. Scanlon (Bloomington, IN: Indiana University Press, 2005).

44. Jeffrey S. Metcalfe, "Hoping without a Future: Augustine's Theological Virtues beyond Melancholia," *Anglican Theological Review* 95, no. 2 (Spring 2013): 242.

45. Thomas Carlson, "With the World at Heart: Reading Cormac McCarthy's *The Road* with Augustine and Heidegger," *Religion and Literature* 39 (2007): 49.

46. Kearnes, "Miraculous Engineering," 225.

47. Brand, "About Long Now."

48. Martin Heidegger, *The Question Concerning Technology and Other Essays,* trans. William Lovitt (London: Harper and Row, 1977).

49. Carlson, "With the World at Heart," 63–64.

50. Ibid., 51.

51. It has also been argued that de-extinction narratives demonstrate quite neatly what Freud had in mind by the pathology of melancholic attachment as the failure of mourning. Emily Thew (2014) makes this point with reference to "self reproaches" that are, in Freud's diagnosis, symptomatic of melancholia; the melancholic effectively suffers a loss of ego itself, but experiences it through a narcissistic identification (of ego) with lost object. De-extinction narratives, Thew argues, appeal precisely to a sense of guilt and self-reproach (over the lost 'collective' ego) transformed into the moralistic duty of scientists to preserve itself (on behalf of all humans) through the 'inconsolable' attachment to lost species.

52. Heather Saul, "Extinct Frog Brought Back to Life," *The Independent* (November 23, 2013).

53. Colm Tóibín, *The Testament of Mary* (New York: Scribner, 2012).

54. See, for instance, Deborah Bird Rose and Thom van Dooren, "Keeping Faith with Death: Mourning and De-Extinction," (2015), http://thomvandooren.org/2013/11/02/keeping-faith-with-death-mourning-and-de-extinction/.

55. In terms of Anthropocene discussions, this is well underway. See, for instance, the special issue "Deconstruction in the Anthropocene," published in the *Oxford Literary Review* 34, issue 2.

56. Bruno Latour, "Facing Gaia: A New Enquiry into Natural Religion," Gifford Lectures 2013. http://www.ed.ac.uk/schools-departments/humanities-soc-sci/news-events/lectures/gifford-lectures/archive/series-2012–2013/bruno-latour.

BIBLIOGRAPHY

Augustine of Hippo. *Confessions*. Translated by H. Chadwick. Oxford: Oxford University Press, 1991.

Brand, Stewart. *The Whole Earth Discipline: Why Dense Cities, Nuclear Power, Transgenic Crops, Restored Wildlands, Radical Science, and Geoengineering Are Necessary*. London: Penguin, 2009.

Brand, Stewart. "About Long Now." http://longnow.org/about/. 2015. Accessed January 9, 2015.

Caputo, John D., and Michael J. Scanlon, eds. *Augustine and Postmodernism: Confession and Circumfession*. Bloomington: Indiana University Press, 2005.

Carlson, Thomas A. "With the World at Heart: Reading Cormac McCarthy's *The Road* with Augustine and Heidegger." *Religion and Literature* 39, 3 (2007): 47–71.

Chrulew, Matthew. "Reversing Extinction: Restoration and Resurrection in the Pleistocene Rewilding Projects." *Humanimalia: A Journal of Human/Animal Interface Studies* 2, 2 (2011): 4–27.

Chrulew, Matthew. "Hunting the Mammoth, Pleistocene to Postmodern." *Journal for Critical Animal Studies*, 9, 1/2 (2011): 32–47.

Church, George M. *Regenesis: Reinventing Nature and Ourselves*. New York: Basic Books, 2012.

Clayton, Jay. "Genome Time." In *Time and the Literary*, edited by Karen Newman, Jay Clayton, and Marianne Hirsch, 31–60. London: Routledge, 2002.

Clingerman, Forrest. "Reading the Book of Nature: A Hermeneutical Account of Nature for Philosophical Theology." *Worldviews* 13 (2009): 72–91.

Clingerman, Forrest. "Geoengineering, Theology, and the Meaning of Being Human." *Zygon*, 49 (2014): 6–21.

Clingerman, Forrest. "Redeeming the Climate." In *Technofutures, Nature and the Sacred*, edited by Celia Deane-Drummond, Sigurd Bergmann, and Bronislaw Szerszyinski, 175–192. London: Ashgate, 2015.

Colebrook, Claire. *Death of the PostHuman: Essays on Extinction*, Vol. 1. London: Open Humanities Press, 2014.

Colebrook, Claire. "Archiviolithic: The Anthropocene and the Hetero-Archive." *Derrida Today*, 7 (2014): 21–43.

Conradie, Ernst. "Resurrection, Finitude, and Ecology." In *Resurrection: Theological and Scientific Assessments,* edited by Michael Welker, Ted Peters, and Robert John Russell, 277–296. Grand Rapids, MI: Eerdmans, 2002.

Daley, B. 2002. "A Hope for Worms: Early Christian Hope." In *Resurrection: Theological and Scientific Assessments,* edited by Michael Welker, Ted Peters, and Robert John Russell, 136–164. Grand Rapids, MI: Eerdmans, 2002.

Derrida, Jacques. "Circumfession." In *Jacques Derrida*, translated by Geoffrey Bennington. Chicago: Chicago University Press, 1993.

Derrida, Jacques. *Of Grammatology*, corrected edition. Translated by Gayatri Spivak. Baltimore: John Hopkins University Press, 1997.

Drassinower, Abraham. *Freud's Theory of Culture: Eros, Loss, and Politics*. London: Rowman and Littlefield, 2003.

Freud, Sigmund. *Standard Edition of the Complete Psychological Works*, Vol. 14. Translated by James Strachey. New York: Norton, 1957.

Galaragga, Maialan, and Bronislaw Szerszynski. "Making Climates: Solar Radiation Management and the Ethics of Fabrication." In *Engineering the Climate: The Ethics of Solar Radiation Management*, edited by Christopher Preston, 221–235. Lanham, MD: Lexington Books, 2012.

Gardiner, Stephen. "Is 'Arming the Future' with Geo-engineering Really the Lesser Evil? Some Doubts about the Ethics of Intentionally Manipulating the Climate System." In *Climate Ethics: Essential Readings,* edited by Stephen Gardiner et al., 284–313. Oxford: Oxford University Press, 2010.

Gaston, Sean. "Derrida and the Eco-Polemicists." *Paragraph* 36 (2013): 344–360.

Guest, Peter. "Giant Cows, Geoengineering and the Lazarus Frog." *Surviving the Future* (2013). https://medium.com/surviving-the-future. Accessed June 3, 2015.
Heidegger, Martin. *The Question Concerning Technology and Other Essays.* Translated by William Lovitt. London: Harper and Row, 1977.
Jantzen, Grace. "Do We Need Immortality?" *Modern Theology* 1 (1984): 33–44.
Kearnes, Matthew. "Miraculous Engineering and the Climate Emergency: Climate Modification as Divine Economy." In *Technofutures, Nature and the Sacred*, edited by Celia Deane-Drummond, Sigurd Bergmann, and Bronislaw Szerszyinski, 213–240. London: Ashgate, 2015.
Klein, Naomi. *This Changes Everything: Capitalism vs the Climate*. London: Allen Lane, 2014.
Latour, Bruno. "War and Peace in an Age of Ecological Conflicts." Lecture at Peter Wall Institute, Vancouver, September 23, 2013. http://www.bruno-latour.fr/node/527. Accessed 25 March 25, 2015.
Latour, Bruno. "Facing Gaia: A New Enquiry into Natural Religion." Gifford Lectures (2013). http://www.ed.ac.uk/schools-departments/humanities-soc-sci/news-events/lectures/gifford-lectures/archive/series-2012–2013/bruno-latour. Accessed November 8, 2015.
Matthews, Steven Paul, "Francis Bacon's Scientific Apocalypse." In *The End That Does: Art, Science, and Millennial Accomplishment,* edited by Cathy Guttierez and Hillel Schwartz, 97–115. London: Equinox, 2006.
Metcalfe, Jeffrey S. "Hoping without a Future: Augustine's Theological Virtues beyond Melancholia." *Anglican Theological Review* 95, 2 (2013): 235–250.
Nightingale, Andrea. *Once Out of Nature: Augustine on Time and the Body*. Chicago: University of Chicago Press, 2011.
Novak, Ben. "Reflections on Martha's Centennial." Revive & Restore (2014). http://longnow.org/revive/category/passenger-pigeon/. Accessed September 23, 2015.
Preston, Christopher J. "Re-thinking the Unthinkable: Environmental Ethics and the Presumptive Argument against Geoengineering." *Environmental Values* 20 (2011): 457–459.
Project Passenger Pigeon. 2015. http://passengerpigeon.org/casestatement.html. Accessed March 4, 2015.
Rose, Deborah, and Thom Van Dooren. "Keeping Faith with Death: Mourning and De-Extinction." http://thomvandooren.org/2013/11/02/keeping-faith-with-death-mourning-and-de-extinction/. Accessed April 29, 2015.
Ruether, Rosemary Radford. *Sexism and God-Talk*. London: SCM Press, 1983.
Ruether, Rosemary Radford. *Introducing Redemption in Christian Feminism.* Sheffield: Sheffield Academic Press, 1998.
Sandler, Ronald. "The Ethics of Reviving Long Extinct Species." *Conservation Biology* 28 (2013): 354–360.
Schildgen, Brenda Deen. "Augustine's Answer to Jacques Derrida in *De Doctrina Christiana.*" *New Literary History* 25 (1994): 383–397.
Shapiro, Beth. *How to Clone a Woolly Mammoth: The Science of De-extinction.* Oxford: Princeton University Press, 2015.
Sherkow, Jacob S., and Henry T. Greely. "What If Extinction Is Not Forever?" *Science* 340, 6128 (2013): 32–33.
Skrimshire, Stefan. "Eschatology." In *Systematic Theology and Climate Change,* edited by Michael Northcott and Peter M. Scott, 157–74. London: Routledge, 2014.
Steffen, Will, Paul J. Crutzen, and John R. McNeill. "Are Humans Now Overwhelming the Great Forces of Nature?" *Ambio* 36, 8 (2007), 614–621.
Steiner, John. "Discussion: Climate Change in a Perverse Culture." In *Engaging with Climate Change: Psychoanalytic and Interdisciplinary Perspectives*, edited by Sally Weintrobe, 81–83. London: Routledge, 2013.
Szerszynski, Bronislaw. "Reading and Writing the Weather: Climate Technics and the Moment of Responsibility." *Theory, Culture and Society* 27, 2–3 (2010): 9–30.
Szerszynski, Bronislaw. "The End of the End of Nature: The Anthropocene and the Fate of the Human." *The Oxford Literary Review* 34, 2 (2012): 165–184.

Thew, Emily. "De-extinction and Melancholia: Narcissistic Attachments." Presentation at Immortality and Infinitude in the Anthropocene, Stockholm, December 2014. http://www.immortalityandinfinitude.net/blog/-emily-thew-university-of-sheffield. Accessed April 28, 2015.

Tóibín, Colm. *The Testament of Mary*. New York: Scribner, 2012.

Turner, Stephanie S. "Open-ended Stories: Extinction Narratives in Genome Time." *Literature and Medicine* 26 (2007): 55–82.

Vessey, Mark. "Reading Like Angels: Derrida and Augustine on the Book (for a History of Literature)." In *Augustine and Postmodernism: Confessions and Circumfession*, edited by John D. Caputo and Michael J. Scanlon, 173–207. Bloomington: Indiana University Press, 2005.

III

Religious Resources for Moral Discernment

Chapter Seven

Healing the Climate? Christian Ethics and Medical Models for Climate Engineering

Laura M. Hartman

When examining an issue such as climate geoengineering, language matters. Like Clingerman and Skrimshire in the previous two chapters of this volume, I am attentive to the significance of metaphors, frames, and models. The very term, "climate geoengineering," contains an assumption that what is going on here is best characterized as "engineering," and by implication that the climate can be helpfully envisioned as a machine to be engineered.[1] In this chapter, I question the use of "engineering" to describe these efforts, because I also question the mechanistic view of the planet. Instead, I offer support for an organic view of the Earth, and a medical model for understanding climate ills and the efforts to remedy them.[2]

Let me begin, first, by acknowledging that a complex phenomenon such as global climate change often requires multiple rhetorical frames in order to fully understand it. We need many perspectives to adequately address such a multifaceted, vexing problem.[3] An organically based medical model is not the only good way to approach the questions of deliberate, large-scale human intervention in the atmosphere intended to ameliorate the bad effects of climate change. I do, however, argue that medical models are better than engineering models, based on criteria I will elaborate below.[4]

ABOUT METAPHORS, MODELS, AND FRAMES

I use two major root metaphors or overarching conceptual frameworks in this chapter: earth as machine and earth as organism.[5] In this chapter I refer to these root metaphors or conceptual frameworks, including the more specific metaphors and frames that arise from them, as models. According to Christian eco-theologian Forrest Clingerman, "a model is a construction that attempts to show how we can cognitively deal with an object": it is a deliberately created heuristic tool.[6] Models, however, also "participate in" and draw upon the unconscious, organic power of metaphors, which reveal unexpected truths by highlighting a previously unnoticed resemblance between two things.[7] The most successful models become "overarching conceptual framework[s]," also known as root metaphors: these are the "big ideas" upon which paradigms are built.[8] For example, the model of earth-as-organism and the model of earth-as-machine underlie two very different worldviews, which lead to very different understandings of climate change and responses to it. Christian theologian Sallie McFague uses the word "model" as a catch-all term to include metaphors, frames, and so forth, whether of recent invention or ancient origin.[9] For her, models draw on metaphors' (somewhat ancient and untamed) power, but they also are subject to the deliberate goals and strategies of those who wield them. We may not completely control our models, and yet we can choose to use some and reject others; we can judge them morally and use them strategically.

In *Metaphors We Live By,* Lakoff and Johnson praise metaphors' conceptual power: "Metaphors may create realities for us, especially social realities."[10] As they see it, "Metaphor is principally a way of conceiving of one thing in terms of another, and its primary function is understanding."[11] The conceptual power of metaphors shapes human understanding of complex phenomena. According to Lakoff and Johnson's paradigmatic example, we talk about arguments as if they were strategic moves in a war, and this shapes our experience of an argument; if we conceived of argument as a dance, we would perceive it differently.[12] As Lakoff and Johnson would agree, metaphors are not only conceptual. In addition to their cognitive content, metaphors can be discursive, defining not just an idea but an entire discourse—as when the "spaceship earth" metaphor enables discussion of "engineering the climate."[13] Additionally, metaphors have a normative dimension. As Dutch environmental philosopher Jozef Keulartz notes, "metaphors enable us—mostly automatically and unconsciously—to make a 'normative leap' from data to recommendations, from facts to values, from 'is' to 'ought.'"[14] If we conceive of the earth as an organism that is ill, for example, then there is an implicit directive to heal it or at least palliate its suffering. The power of metaphors, then, is cognitive, discursive, and normative.

British linguist Brigitte Nerlich and psychologist Rusi Jaspal acknowledge the multiple powers of metaphor by casting metaphor itself as a type of technology: "In some respects metaphor, a tool we use to think and act with, is a linguistic technology that needs as much ethical oversight as the technologies we 'see through' it, such as geoengineering."[15] In so doing they are highlighting the fact that metaphors are manufactured by humans for a purpose, and it is worthwhile to uncover these purposes, and their effects, to evaluate the usefulness of the metaphor. Nerlich and Jaspal rightly understand that metaphors may be created deliberately; but metaphors also arise unconsciously, organically, in human cultures.

Nerlich and Jaspal describe metaphors as narratives: "short stories that invite interpretation and action."[16] A metaphor tells a story with a normative dimension, describing what is and setting us up to recognize what ought to be. Climate communications researcher Matthew Nisbet says much the same thing of "frames," which he defines as "interpretive storylines that set a specific train of thought in motion, communicating why an issue might be a problem, who or what might be responsible for it, and what should be done about it."[17] According to Nisbet, situations that are "ambiguous" or "uncertain"—such as the complex, challenging problem of climate change—necessitate the use of frames. He writes, "For many members of the public, climate change is likely to be the ultimate ambiguous situation given its complexity and perceived uncertainty."[18] Frames help us think: like metaphors, they are cognitive, discursive, and normative. Nisbet writes, "A frame links two concepts, so that after exposure to this linkage, the intended audience now accepts the concepts' connection."[19] Frames are metaphors writ large; they tell the audience what is true and real, and what should be done.

Drawing on these insights, in this chapter I will follow McFague to use the word "model" inclusively, assuming that metaphors and frames are types of models with which people understand and orient themselves toward the world.

FRAMING GEOENGINEERING

The earth-as-machine model is a powerful one in geoengineering discourse. Nerlich and Jaspal argue that geoengineering has been framed almost exclusively in technological terms: the earth is a machine that requires fixing. Hence, discussions of geoengineering feature images of a thermostat and a sunshade or sunscreen; language of toolboxes, toolkits, and tinkering; and most recently, "hacking," which compares the atmosphere to a computer.[20] Climate geographers Kate Porter and Mike Hulme found a distinct bias in favor of technological terms in their frame-based analysis of British media representations of geoengineering. They write, "Debate on geoengineering in the British national press has thus far been centered on an understanding of geoengineering as a technological innovation and as an issue of risk management."[21] Porter and Hulme observe that the Royal Society leads and shapes the discourse on this topic, and they are "almost inherently pro-technology."[22] This bias seems not to be limited to the British press; American discussions of geoengineering follow a similar mold.

The earth-as-organism model, by contrast, sees the Earth and living organisms not as machines to be tinkered with; they are holistic, living entities. Many thinkers have seen the earth as an organism, from Scottish philosopher David Hume to earth systems scientist James Lovelock, and from indigenous traditions to contemporary Christian theologian Sallie McFague.[23] According to historian Carolyn Merchant, in sixteenth-century Europe and earlier, "the root metaphor binding together the self, society, and the cosmos was that of an organism."[24] But the European Enlightenment era led to a paradigm shift: organisms, and indeed the earth itself, were seen as machines. Intellectual discourse on nature reached the conclusion "that it is not merely useful, under certain circumstances, to regard organisms *as if* they were machines, but rather that it is *only* by conceiving organisms *as* machines that we can truly make sense of them."[25] The root metaphor of earth-as-machine took hold in European thought. Now, it dominates Western scientific discourse. In McFague's words, "We believe in, live in, the machine model as fishes live in the sea; it is not to most of us a construct, but the way things are."[26] This has implications for the way we think about humans and nature. McFague writes, "It is a sensibility that above all else assumes the primacy of rational control: it is, in significant ways, the apotheosis of the human—over self, over nature, over God."[27]

What about an organismic view of the earth? Here we come into another metaphor: James Lovelock's Gaia theory. This idea, that the earth is a living organism, is based on a root metaphor that would be recognizable to Merchant's sixteenth-century peasants and to indigenous people around the world—earth as the body of a god or goddess. To scientists this may sound frivolous, but Gaia theory itself is meant to be thoroughly scientific, positing that Earth is, in Lovelock's words, "a self-regulating system made up from the totality of organisms, the surface rocks, the ocean and the atmosphere tightly coupled as an evolving system. The theory sees this system as having a goal—the regulation of surface conditions so as always to be as favorable as possible for contemporary life."[28] Self-regulating, evolving, goal-directed: these describe an organism more clearly than they describe a machine. In terms of earth systems science, the planet as a whole may be fruitfully compared to a living organism.

Truthfully, the Earth as a whole is neither a machine nor an organism: it is a unique entity without clear parallel in the known universe. But the question is one of models: How should these models be judged? And is it even legitimate to do so? Metaphors, models, and frames are both consciously chosen and unconsciously inherited; they grow both deliberately and organically. Insofar as they are consciously chosen and deliberately cultivated, we may judge them. I propose to judge based on truthfulness, usefulness, fruitfulness, and inclusivity. As Lakoff and Johnson recognize, metaphors selectively direct attention, and in so doing they both highlight some true things and hide other true things.[29] Analyzing a model involves first uncovering what is highlighted by the model and what is hidden, and then assessing both the truthfulness and the usefulness of these highlighted and hidden aspects. Is it fruitful? What fruits does it bear and how do we assess them? Models may prove to be more or less helpful in the quest to support the well-being of non-human nature, if they do or do not promote attitudes and values that are likely to result in beneficial human actions; those that are more helpful are preferred in this context. Finally, all other things being equal, I am in favor of models that include more players in the conversation, since inclusivity and participation are positive values.

First let us discuss the machine model for nature, and then the machine model for the planet as a whole (particularly the atmosphere, since that is the entity that might be "engineered" in this scenario). The machine model for nature highlights the intricacy, intelligibility, and cause-effect relationships in the natural world. It hides nature's holism, ineffability, and unpredictable dimensions. If we measure models in part by their fruitfulness, then let it be noted: the machine model (mechanistic frame) has led to very positive, productive developments in Western science. Thanks to this view, scientists have come to understand the "functioning" of different components of the cell; the "purpose" of different organs in a body; the "logic"—perhaps even "design"?—behind an organism's role in an ecosystem.[30] We can scarcely imagine what life science would be like without this major metaphor for the understanding of living organisms and life systems.

However, as philosopher of science Jordan Nicholson argues, in the end, living beings are not the same as machines. The model is not fully truthful, and this curtails its usefulness. Unlike machines, living beings are best understood as organisms: self-organizing, self-repairing, self-controlling, and self-directing. Organisms metabolize and reproduce in ways drastically different from machines. Machines are designed to fulfill a human designer's purpose; organisms evolve without a similar human-directed purpose.[31] To "inadvertently mislead non-specialists into assuming that organisms *really are* machines" causes great harm, according to Nicholson.[32] He does not venture into the philosophical question of intrinsic value, but clear arguments can be made that a living being has more claim to intrinsic value than a machine, due to some of the factors Nicholson mentions—the living thing possesses a greater degree of selfhood, and it is intrinsically purposive rather than extrinsically purposive.

While Nicholson discusses organisms and biology, this machine model can also apply to the atmosphere, and to the Earth as a whole. Complex and chaotic though it may be, in the eyes of geoengineering advocates, the atmosphere is something that can be "tinkered with," or "hacked" to optimize a benign climate for human and other life. The earth's atmosphere does resemble a machine that might be engineered: it is made up of multiple parts; the parts are nonliving (though they do interact with living things); despite the chaotic factors, there are certain elements of cause-and-effect in the atmosphere that can be modeled, charted, and perhaps controlled. The machine model highlights these aspects of the truth. It obscures other aspects: the atmosphere is *not* a machine insofar as it was not designed by a human to fulfill a particular purpose; its physical makeup is much more fluid and changeable (e.g., the flow of oxygen into plants or the evaporation of water into the clouds) than the physical makeup of most machines; it is much more "wild" and hard to predict than the behavior of most machines. Lovelock would add that the atmosphere is a component of a larger system, Gaia, which can be called a living being. The atmosphere, then, is part of something that is self-regulating, evolving, and intrinsically purposive: the machine conception obscures these truths.

What are the fruits of the machine model for the planet? To some extent, discussions of "spaceship earth" can inspire eco-friendly attitudes of restraint and prudence: useful for those seeking to conserve Earth's resources. However, McFague's note about arrogance, above, indicts the machine model for fostering human attitudes of hubris and control that are less conducive to environmentally sound behavior. Would-be geoengineers certainly can be accused of arrogance (see O'Brien's chapter in this book, in which he argues that the humble act of repentance is a prerequisite for even considering geoengineering). Finally, engineering metaphors tend to be exclusionary. While many humans are somewhat empowered to create and tinker with our own machines, most of us live in a world where engineers are a small elite class who know and understand mechanical complexity. There is, additionally, a gender dimension to our society's expectations about who can and should be engineering machines, which adds to the exclusivity of this model.

Overall, then, the machine model obscures important, salient dimensions of the nonhuman world and the planet as a whole; it bears bountiful fruit in terms of an understanding of the natural world; but it also risks arrogance and exclusion in the actions that it fosters.

THE MEDICAL MODEL

Lovelock clearly prefers a medical metaphor when discussing Gaia, the atmosphere, and climate change. He writes, "I speak as a planetary physician whose patient, the living Earth, complains of fever; I see the Earth's declining health as our most important concern."[33] Throughout *The Revenge of Gaia* he elaborates on this metaphor, describing farming as abrasive to Gaia's "skin" or the Earth as "elderly and less resilient" than in ages past.[34] Lovelock in this book compares efforts at climate geoengineering to "the use by paramedics of oxygen for heart failure and breathing difficulty or a pressure pad for a hemorrhage—something temporary, to keep a patient alive until they reach the full services of a hospital."[35] Lovelock is so deeply alarmed by the state of Gaia's health (in terms of climate change, primarily) that he supports geoengineering as a method to "buy us time."[36]

Lovelock recognizes that certain forms of geoengineering enslave us to perpetual maintenance of the atmosphere. "The more we meddle with the Earth's composition and try to fix its climate, the more we take on the responsibility for keeping the Earth a fit place for life, until eventually our whole lives may be spent in drudgery doing the tasks that previously Gaia had freely done for over three billion years."[37] Such drudgery may be the price we pay for the time that geoengineering "buys" us—well worth it if it extends human-Gaia life and moves us toward symbiosis with, rather than parasitism upon, the living planet.[38] Lovelock believes that geoengineering efforts, however, should also be accompanied by "something like the Hippocratic Oath that physicians utter: 'Do nothing that would harm the patient.'"[39] Geoengineering, then, is tantamount to a drastic medical measure, taken in the name of life-support in an extreme situation. And the duty to nonmaleficence—not harming the living planet—remains intact.

When it comes to understanding climate change and inspiring action to help ameliorate it, imagery involving health has great appeal. O'Brien puts it well: "A health-focused metaphor would appeal to the universal experience of illness and the uncomplicated moral claim that health is a good that should be protected and restored whenever possible."[40] Furthermore, a health metaphor "provides a vocabulary of symptoms, syndromes, diagnostic indicators, and so on with which laypeople are already familiar," Keulartz notes.[41]

There is also evidence that health-focused framing on climate change is effective. According to Nisbet and his collaborators, if the goal in communicating about climate change is feelings of hope and efficacy—feelings that typically lead to engagement and action—then some frames work better than others. According to a 2012 study, gloom-and-doom scenarios typically "result in *less* concern and *more* hopelessness among audiences."[42] Even a national security frame—emphasizing the international military risks associated with climate change—did not create hope and engagement, except among those already alarmed about climate change. The best frame to generate hope and engagement, according to this study, was a health frame: describing climate change as an issue of human health.[43] This is not exactly the same as using a metaphor of the planet as a sick body, but it does support

O'Brien's and Keulartz's positions, that health framing is more relatable and actionable than some of the alternatives.

Lovelock is not the only advocate of geoengineering who uses a health model to support these efforts to alter the Earth's climate. Stanford environmental scientist Stephen Schneider asserts that our society is addicted to fossil fuels and needs to go into rehab; perhaps geoengineering is our methadone or nicotine patch.[44] Harvard environmental scientist David Keith has compared geoengineering to chemotherapy—not a treatment that is desirable in and of itself, but when one's health condition is truly dire, it is better than the alternative.[45]

The health model does have its critics, however. Nerlich and Jaspal criticize the health model when it is used to support climate geoengineering. "Healing and medicine," they observe, "are directly linked to illness, medical failure and death as well as medical quackery. Healing or restorative metaphors for geoengineering can therefore be easily subverted by saying, for example, that geoengineering is like liposuction for the planet."[46] These factors, however, do not mean that using medical models for geoengineering is wrong: while these models may be used to justify geoengineering, they may also be used to rightly criticize it.

Before we can evaluate the efficacy of the medical or health model for climate problems, we must explore more fully several different understandings of medicine in contemporary Western cultures. Medicine is a science, but it also has spiritual overtones: Jesus and his disciples healed the sick, effecting miraculous cures for medical outcasts in their time. Christians have built hospitals, trained medical professionals, and frequently viewed medicine as a kind of ministry.[47] Christians have a reason to prefer the medical model for use in discussing climate change: for better or worse, the tradition contains a much richer discussion of the spirituality and ethics of healing than of engineering.

Of course, medicine is also a science. As with all sciences, medical science has been subject to a mechanistic root metaphor, and yet its subject matter is nonmechanized living beings. Some medical interventions are more organic in nature, and others are more like medical engineering. There are multiple types of medicine available to patients in the industrialized world. Many of us partake of Western or biomedicine (vaccines, surgeries, chemotherapy, and most everything else in a typical hospital); Eastern medicine (such as acupuncture, which is based on an understanding of the balance or imbalance of ineffable energy meridians in the human body); folk medicine (as when I stay warm to avoid catching a "cold"); herbal medicine (the Echinacea tea I drink when I get the cold); psychological medicine (in which I make myself feel better physically by reaching out to a loved one emotionally); and of course preventive medicine (such as getting a good balance of exercise and rest, which every form of medicine sees as healthy). This panoply of options within the medical model is a strength for two reasons. First, if we see the earth as a sick organism, it is not immediately obvious that "surgery" is necessary; perhaps the earth needs "rest" or "acupuncture" (depending on the diagnosis). Second, while Western medicine (and to some extent Eastern medicine) is indeed the domain of well-educated elites, these other forms of medicine are quite accessible to laypeople as practitioners. Both of these mean that a medical model will yield more options, and more accessible conversations, than the machine model.

Those who support geoengineering tend to compare it to more high-tech interventions such as oxygen masks and chemotherapy (see Lovelock and Keith, above). Those who are against geoengineering either compare it with frivolously high-tech medical procedures ("liposuction" according to Nerlich and Jaspal, above) or prefer a metaphor based on alternative medicine or Eastern medicine (as when reforestation activist Andy Lipkis describes planting trees as performing "acupuncture for the planet"—this may be taken to imply that Western high-tech interventions are less effective).[48]

Within Western medicine, there is an internal debate about the usefulness of a machine metaphor for the body. Medicine, to many, seems impersonal and too mechanized. According to philosopher of medicine Drew Leder, this can be traced to the model of "Man-the-Machine," which includes reducing every biological phenomenon to its constitutive chemistry and physics.[49] Leder critiques this "Cartesian model of embodiment" for its inherent dualism between soul and body, supporting alternative feminist and phenomenological models of the body.[50] Philosopher of science Keekok Lee takes this argument even farther. She contends that the Cartesian machine model of organisms was not simply a new metaphor but a revolutionary worldview, "an ontological *volte-face*," with "widespread implications" throughout culture and society.[51] Machines are human artifacts, "the material embodiment of human intentionality."[52] To see organisms, including human bodies, in this light means reductionism and instrumentalism: medicine becomes a form of engineering, rather than, say, teaching, mentoring, creating, nurturing, etc.[53] But experience shows that human bodies are not machines; the intricacies of human health do not line up in the discrete cause-and-effect pattern that the machine model would suggest (Lee mentions the Placebo effect as an example).[54] When I invoke a medical model for discussing the earth, then, I wish to bring with it the complexity implied by these thinkers. We know from our own experience that there is more to human health, and more to the health of the planet, than well-oiled gears. Ultimately, we seek healing and wholeness, these holistic and organic concepts that are not adequately captured by a machine model.

Let us then evaluate the medical model: the Earth is a large organism, and it is sick. What does this model highlight and what does it hide? It highlights important aspects of the climate problem: it is complex, systemic, and involves deeply interconnected systems; it is a matter of life and death (at least, to humans and many other life-forms if not to the earth itself); treatment that addresses symptoms rather than the disease itself may prove ineffective; and though we have healing powers at our disposal, the ultimate outcome is uncertain. But this metaphor also hides certain aspects of the climate problem: our knowledge of climate is much less certain than our medical knowledge of the human body; there is only one Earth, which means that we cannot "experiment" on a solution with other organisms before administering it to our planet; and to pinpoint one analogous malady for the Earth (e.g., "the Earth has HIV!") might oversimplify the situation to the extent that it rings false.

To continue the evaluation: what are the fruits of this model and what attitudes and behaviors would it encourage or inhibit? The medical model yields important conversations, in echo of some conversations within medicine: how mechanized is the planet, and how much technology should we use in our responses to its ailments? How much control can we really have over this particular "patient"? Furthermore, how should we best characterize Earth's illness? What "treatments" are most appropriate? What is the role of "preventive medicine" or "alternative medicine" or "palliative care"? Who among us are "doctors" or "acupuncturists" or "pharmacists" or "candy stripers" or "hospice workers"? This points to the inclusivity of this model. While few of us are paid medical professionals, we all have bodies and we all care for the health of our bodies; it is a model that a broad range of people can understand and relate to.

On the negative side, the medical model also generates defeatist perspectives. If the earth is a sick organism, perhaps the illness is humanity. Seeing humans as a disease or a parasite is not an inaccurate comparison, perhaps, but one that may have deleterious effects if it leads to despair and misanthropy. Views of the earth as a doomed patient raise other questions: should we just pull the plug? Put the planet on hospice care? Should we live it up now and let the planet's health deteriorate later, shrugging that "you only live once"? These concerns point to another weakness of this metaphor: the planet is a collective entity, not a singular entity. To pursue planetary "health" requires cooperation between many, many people, whereas personal health is more singular. Furthermore, one human's "good death" after a long and useful life is not a great tragedy, but the death of the planet is entirely different. Of course, though the global situation is dire, no scientists seriously anticipate the imminent death of the planet—though the survival of humans and other more vulnerable species is certainly at risk—so this, too, is a confusing inaccuracy of the medical model.

It may not be perfect, but I rate the medical model highly. I think that it highlights important and useful aspects of the global situation; it is fruitful and generative in positive directions that, in my view, outweigh the negative potential; and its inclusivity and relatability are positive.

COMPARISON WITH DIABETES

Perhaps the most helpful aspect of the medical metaphor is that it allows us to evaluate geoengineering from a new perspective. As the multiple medical perspectives already noted indicate, it is fruitful to ask: What is climate change? Is it an addiction, a cancer, an organ failure? And what *is* geoengineering? Is it a nicotine patch, chemotherapy, an oxygen mask? These questions, of course, are about the nature of the Earth's malady as well as the proposed treatments.

To name one disease from which the Earth suffers is oversimple, and to string along an awkward comparison can be obtuse. Nevertheless, in the remainder of this chapter, I explore a particular form of the medical model for climate change because of its interesting implications for the larger discussion of models. Inspired by the work of Australian health scholar Garry Egger, I explore the model of our climate-changed earth as a prediabetic (type 2) patient.[55] Diabetes is a disease constantly being researched, but the current state of scientific understanding is as follows: Normally, when a person consumes carbohydrates and they enter the bloodstream, the hormone insulin functions to help cells use the carbohydrates (glucose) for energy. For type 2 diabetes, an individual who has certain risk factors (sedentary lifestyle, obesity) may become prediabetic, a situation characterized by an increasing inability to metabolize carbohydrates. The levels of glucose in the blood is elevated; either insulin is inadequate, or the cells have become insulin-resistant. If this continues beyond a certain level, the individual is declared diabetic. Excess glucose in the blood is characterized, at first, by thirst, excess urination, and hunger. Later stages of untreated diabetes can lead to heart disease, strokes, problems with the retina leading to blindness, kidney failure requiring dialysis, and poor blood flow in the limbs requiring amputations. Ultimately, untreated diabetes can be fatal.

Type 2 diabetes can be prevented, or its onset can be delayed, by reducing the major risk factors such as obesity and sedentary lifestyle.[56] Patients in the prediabetic state are encouraged to alter their lifestyles to ward off the disease. This often entails losing weight, exercising, and eating healthful foods. Even patients with a recent diagnosis of diabetes can reverse the diagnosis through such preventive measures. If the disease has progressed too far, however, or if the patient is unable to complete these lifestyle alterations,[57] then treatments such as insulin injections (to augment the body's inability to produce or use it) and kidney dialysis (due to diabetes-related kidney failure) may be necessary. Diabetes is a fairly complex disease, but it is also quite commonplace, so most people in the developed world are familiar with it. Climate change, also, is complex but (should be) widely familiar.

The similarities between diabetes and climate change are fascinating. Egger believes that obesity (one of the risk factors associated with type 2 diabetes) can be understood as a metabolic "inflammation," and that the warmed climate is also, in some sense, "inflamed."[58] He sees the Earth as a prediabetic, obese patient. Egger writes, "The warning sign—the 'insulin resistance" of the ecosphere—is illustrated in the inability of the planet's sinks to 'soak up' excessive carbon emissions ('carbon resistance'?), just as the body can no longer effectively utilize blood sugars because of the failure of its cellular 'sinks' through insulin resistance."[59] In both cases—individual prediabetes and planetary warming—the common problem is "the growing misuse of energy; too little personal energy expended and too much high-energy dense food consumed in the case of the individual, and too much fossil fuel energy burned and waste emissions given off in the case of the broader environment."[60] The prescriptions for a prediabetic patient (lose weight, exercise more, eat healthfully) all pertain to the use of energy, either stored as fat, ingested as food, or expended as exercise. So with our metaphorically prediabetic planet: we need to use less fossil fuel energy and store more carbon (e.g., by planting more trees, or through carbon capture and storage technologies).

If efforts in this direction fail, of course, the patient faces ever more serious health consequences. Perhaps we see this in the Earth already. Are extinct species the planet's blinded eyes? Are disappearing ice caps and glaciers the planet's amputated limbs? Are the acidified oceans, no longer able to absorb any more carbon dioxide, the planet's failed kidneys—and if so, is an artificial attempt to reduce the effects of warming when the oceans can no longer absorb the impact, a kind of kidney dialysis? Kidney dialysis involves the patient being attached to a machine to clean the blood daily, or sometimes more frequently. It is, in a sense, an artificial kidney—one of the many triumphs of mechanized medicine that keeps patients alive in dire circumstances.

The diabetes metaphor is more accurate than many other healing-geoengineering metaphors. Lovelock, for example, names geoengineering as a drastic, temporary medical measure (an oxygen mask or a pressure pad administered by paramedics), but he admits that humans would need to maintain such measures in the long term: rather than paramedics, we need full-time nursing care. Indeed, as with diabetes, the situation comprises a chronic condition that has no clear cure, though it can be prevented by undergoing challenging lifestyle changes. Unlike Keith's chemotherapy metaphor, however, diabetes points to clearer causes for the malady. Cancer, by and large, is caused more by luck than by lifestyle (with a few exceptions). Type 2 diabetes, however, is associated with certain risk factors that are (theoretically) within an individual's control: weight and activity level. This is more accurate to the situation of our climate: it is a slowly growing problem, we see the risk factors, and they are within our control, yet we do not take action to change them. The nongenetic risk factors for diabetes are largely controllable; similarly, humans can control the amount of carbon dioxide we emit and we can prevent ourselves from deforestation and other blows to the earth's carbon sinks.

I want to pause here and describe my hesitation in using this metaphor. It is far too easy to blame individuals with type 2 diabetes for their condition, particularly because it is linked with obesity. In our industrialized, food-obsessed culture, we tend to vilify people who are overweight, blaming them for their condition and judging them for lacking the virtue to be thin. In a culture that is saturated with unhealthy foods and multiple technologies that enable a sedentary lifestyle (from automobiles to televisions to Internet shopping), it is disingenuous to expect all people to be able to control their own weight. (And of course, for many, there are genetic factors that contribute to body shape.) I certainly do not want to join the "fat shaming" chorus, nor do I wish to blame the sick for their illness.

Nevertheless, the risk factors—diet, weight, and exercise—are within human control, especially in a context where public health measures facilitate the needed healthy behaviors. I wish to emphasize a public health approach to a problem like diabetes because such an approach avoids "fat shaming" and most readily contributes to a constructive discussion of climate change. Public health is an approach to health that is focused on prevention, works at the social more than at the individual level (e.g., by changing cultural assumptions about healthy behaviors), and is focused on the environmental factors that contribute to diseases.[61] Theologians have argued that Christians should promote and be deeply involved in efforts supporting public health, because compassion, justice, and care for the worst-off call for it.[62]

The public health solution to obesity involves multiple factors: access to affordable healthful food, walkable neighborhoods, employment that is not exclusively sedentary, active commutes, green spaces for safe recreation, and cultural shifts in favor of healthy habits.[63] Arguably, climate change is also a "public health" problem, to which the solution is not preaching virtue to potential "green consumers" (who, when they fail, feel a similar shame to the "fat-shamed" obese individuals; see Fredericks's chapter in this volume for a fuller discussion of shame). The public health solution to climate change is parallel: changing the cultural conditions so that individuals are enabled to consume lower amounts of fossil fuels. Interestingly, certain measures such as walkable neighborhoods and less-processed foods are common solutions to both problems.[64]

The strength of the diabetes metaphor is the attention it draws to the range of appropriate responses to the problem. It is clear that the "lifestyle change" responses to the Earth's prediabetes diagnosis are saner, wiser, and less costly than anything like kidney dialysis. To discount diet and exercise as solutions, and to rely only on the prospect of using kidney dialysis to keep one alive, would be an example of Stephen Gardiner's "creative myopia"; in other words, it shows how ludicrous is the choice to rely on geoengineering as a solution to climate change, when the clear remedy is reducing carbon emissions.[65] The specter of kidney failure—and blindness, and limb loss, and eventual death—should prompt the prediabetic patient to take whatever measures are necessary to prevent the disease from progressing. The required lifestyle changes to prevent diabetes are very challenging, but not impossible: a life of more physical activity and a constrained diet is, by most measures, better than a life on dialysis. Though a life on dialysis is possible, it is hobbled and shortened; it is no bright future, but a constant "drudgery" (in Lovelock's words).[66] Even those who receive a transplanted kidney—the hope of most patients on kidney dialysis—while they are grateful to be alive at all, must still submit to a regime of medications and the fear that the new organ may not incorporate well with the old. Without discounting the wonders of organ transplantation, it seems safe to say that a life of exercise and healthy eating is much more desirable than life with a transplanted organ.

Of course, the line between knowing what I ought to do and actually doing it is neither clear nor short. As Fredericks points out in her chapter in this volume, humans often fail to do what they know to be right. This chapter cannot fully address that gap between "should" and "did." But for the case of climate engineering, when humans in groups are deliberating what action comes next, this metaphor should function cognitively, and perhaps also emotionally, as a motivating factor. Failing to reduce carbon emissions and enhance carbon sinks is like signing up for severe, untreated diabetes. Geoengineering is not a technical fix for a malfunctioning "spaceship earth"; it is the frightening prospect of a patient whose disease is progressing toward a stage that requires daily reliance on a machine simply to stay alive.

Stephen Gardiner is concerned that some forms of geoengineering are "shadow solutions": they solve part of a problem but purport to solve the whole problem.[67] Solar radiation management, for example, might cool the globe, but it does not address ocean acidification due to excess carbon dioxide in the atmosphere. The diabetes metaphor avoids this problem: in no way is kidney dialysis a solution to the problem of diabetes. It is a partial fix for one symptom; it does not regulate the levels of blood glucose or the efficacy of insulin; and it cannot be seen as a solution to the entire disease. Similarly, solar radiation management measures do not cure the ills of climate change—not by a long stretch—but they might keep us alive a bit longer if the situation requires it.

This metaphor does not preclude research into geoengineering. Dialysis machines, imperfect though they are, do help extend lives, and those lives do have value. The staunchest advocates of public health and preventive cures to diabetes should support the development of kidney dialysis machines, even if such machines were only intended to help patients with preventable diabetes (which, of course, they are not—many conditions may cause kidney failure). There is concern that simply undertaking research into geoengineering would create a "moral hazard"—humans would be less motivated to undertake the difficult tasks of carbon reduction if this easy "technological fix" were available. But dialysis machines do not (as far as I know) function as a "moral hazard" for prediabetic patients. Quite the contrary: they function as a frightening, "nearly worst-case scenario" that spurs the patient to take immediate preventive measures.[68] If we firmly understood geoengineering to be an uncomfortable, burdensome, last-ditch measure to preserve life—as kidney dialysis might be understood—then the moral hazard loses some of its worry. Research into geoengineering is less morally troubling if we recognize that such measures are truly "*undesirable . . . ungovernable . . .* [and] *unreliable,*" in Mike Hulme's words, even though in a worst-case scenario they may extend human habitability on the planet.[69]

Rather than posing a moral hazard, research into geoengineering within a context of the medical metaphor invites conversation about quality of life and "good death"—to what degree do you or I want our lives extended by the use of machines (now referring to breathing machines, feeding tubes, and so forth)? To what extent should we keep the earth artificially fit for human life? There are no clear answers here, but the questions prompted by the medical metaphor put us in the right context: we are talking about living things, about life stories and appropriateness, not about a "spaceship earth" that needs fixing.

CONCLUSION

Which ideas or images are the most appropriate for climate change and geoengineering? The way metaphors function in ecological restoration discourse is illuminating. Jozef Keulartz notes that in ecological restoration discourse, less degraded ecosystems are treated with aesthetic metaphors (for example, restoring the beauty of pristine wilderness), while more degraded ecosystems are treated with engineering and cybernetic metaphors (reconstructing the ecosystem, hacking the brownfield's soil microbiome). Areas that fall in the middle of the degradation spectrum are typically discussed with medical metaphors.[70] This raises a question: how degraded is Earth's atmosphere? Do we see it as a wilderness requiring aesthetic restoration, or as a brownfield requiring hacking? Opponents of the engineering view of the climate are really pushing back against a view of the atmosphere as degraded beyond hope of natural remedies. I agree. We need not "hack" the atmosphere yet, but neither can we advance a solely aesthetic view of the task before us: medical models strike the right tone between preservation and engineering. Medical models are some of the most useful tools we have for understanding climate change, and they should be used to evaluate the morality of geoengineering.

Clive Hamilton argues that "technological thinking structures our consciousness in a thousand subtle ways that make climate engineering attractive, indeed, almost inevitable."[71] The technological, mechanistic frame is pervasive, not only in science but in society at large. However, this is not the only way to view the planet. Indeed, even the science of ecology "has roots in organicism—the idea that the cosmos is an organic entity, growing and developing from within, in an integrated unity of structure and function."[72]

The organismic root metaphor for understanding the planet, and the medical model for understanding climate change, are truthful and fruitful. They are not perfect, but they offer a strong contribution to the discussion. They invite wide-ranging debates about interconnected systems, alternative therapies, quality of life, and the proper role of humans. The call to heal illness is timeless, and inspires Christians and others to bring spiritual wellness and wholeness to situations of imbalance and fragmentation that manifest in illness. Medicine invites (or should invite, at least) greater humility than engineering: we are working with ineffable organic systems, with complexity beyond human comprehension, and ultimately with a telos that is beyond our control. Humans have many healing powers, but we cannot completely control the outcome of medical interventions. We work with God, the ultimate healer, to harness an organism's self-healing properties; we do not effect healing alone. This humility is a much needed antidote to an engineering perspective characterized by hubris and control.

NOTES

1. Geoengineering is not the only area of study that presumes the planet is a machine to be engineered. The emerging field of "Earth Systems Engineering and Management" (which does tend to be supportive of geoengineering) also presupposes that the planet is helpfully viewed as a machine.

2. For simplicity's sake I will use the term "geoengineering" or "climate geoengineering" to refer to large-scale technological efforts to counteract some or all of the effects of climate change, even though we might conceive of these efforts as "geomedicine" or "geo-life-support."

3. When I use the word "we" in this chapter, I am usually referring to educated, well-off people in developed nations. I choose this audience because people who fit that description are the primary perpetrators of the climate chaos that geoengineering seeks to redress, and we are also the people most likely to execute a geoengineering scheme. Therefore, we need to deliberate carefully about this topic, and I hope this chapter serves that end. Our deliberation should also include listening to voices from less-educated, less-well-off people and from nonhumans; unfortunately, this chapter does not fulfill that need.

4. I recognize that medical interventions in humans are often described using engineering models, as if the human body were a machine rather than an organism. I will address this below.

5. Merchant uses "root metaphor," and Clingerman uses "overarching conceptual framework"—though they sound different, I believe both are gesturing to the same thing, the underpinnings of what is often called a "paradigm" or set of assumptions making up a worldview. See Carolyn Merchant, *The Death of Nature: Women, Ecology, and the Scientific Revolution* (New York: Harper and Row, 1980), 1; and Forrest Clingerman, "Seeking the Depth of Nature in a Scientific World," in *Technology, Trust, and Religion: Roles of Religions in Controversies on Ecology and the Modification of Life*, ed. William Drees (Leiden University Press, 2009), 145.

6. Clingerman, 145.

7. Ibid., 144.

8. Ibid., 145.

9. According to McFague, models are metaphors, frames, and narratives with "staying power," large and stable enough to "present a pattern for relatively comprehensive and coherent explanation." Sallie McFague, *Models of God: Theology for an Ecological, Nuclear Age* (Minneapolis: Fortress Press, 1987), 34.

10. George Lakoff and Mark Johnson, *Metaphors We Live By* (University of Chicago Press, 1980), 156.

11. Ibid., 36.

12. Ibid., 4–5.

13. Jozef Keulartz, "Using Metaphors in Restoring Nature," *Nature and Culture* 2, 1 (Spring 2007): 27–28.

14. Ibid., 28.

15. Brigitte Nerlich and Rusi Jaspal, "Metaphors We Die By? Geoengineering, Metaphors, and the Argument from Catastrophe," *Metaphor and Symbol* 27 (2012): 133.

16. *Ibid*, 133.

17. Matthew C. Nisbet, "Communicating Climate Change: Why Frames Matter for Public Engagement," *Environment* 51, 2 (March 1, 2009): 15.

18. Ibid., 16.

19. Ibid., 17.

20. Nerlich and Jaspal, "Metaphors We Die By?" 135–138. See also Eli Kintisch, *Hack the Planet: Science's Best Hope—or Worst Nightmare—for Averting Climate Catastrophe* (New York: Wiley, 2010).

21. Kate Elizabeth Porter and Mike Hulme, "The Emergence of the Geoengineering Debate in the UK Print Media: A Frame Analysis," *The Geographical Journal* 179, 4 (December 2013): 345.

22. Ibid.

23. See Maarten Boudry and Massimo Pigliucci, "The Mismeasure of Machine: Synthetic Biology and the Trouble with Engineering Metaphors," *Studies in History and Philosophy of Biological and Biomedical Sciences* 44 (2013); James Lovelock, *The Revenge of Gaia: Earth's Climate Crisis and the Fate of Humanity* (New York: Basic Books, 2006); Carolyn Merchant, *The Death of Nature*; and Sallie McFague, *The Body of God: An Ecological Theology* (Minneapolis: Augsburg Fortress, 1993).

24. Merchant, *Death of Nature*, 1.

25. Daniel J. Nicholson, "Organisms ≠ Machines," *Studies in History and Philosophy of Biological and Biomedical Sciences* 44 (2013): 670.

26. McFague, *Body of God*, 34.

27. Ibid.

28. Lovelock, *The Revenge of Gaia*, 162.

29. Lakoff and Johnson, *Metaphors We Live By*, 10.

30. Nicholson, "Organisms ≠ Machines," 669–670. Nicholson recognizes that "machine" language from scientific literature gives fuel to advocates of "intelligent design" (676). (See also Boudry and Pigliucci.)

31. Nicholson, "Organisms ≠ Machines," 671–674.

32. Ibid., 675.
33. Lovelock, *The Revenge of Gaia*, 1.
34. Ibid., 2, 132
35. Ibid., 152.
36. Ibid.
37. Ibid.
38. For more on "buying time," see Clingerman's chapter in this volume.
39. Lovelock, 132.
40. Kevin J. O'Brien, "The 'War' against Climate Change and Christian Eco-Justice: Ethical Implications of Martial Rhetoric," *Worldviews* 16 (2012): 135–153, 150.
41. Keulartz, "Using Metaphors in Restoring Nature," 37.
42. Teresa A. Myers, Matthew C. Nisbet, Edward W. Maibach, and Anthony A. Leiserowitz, "A Public Health Frame Arouses Hopeful Emotions about Climate Change," *Climatic Change* 113 (2012): 1107.
43. Ibid.
44. Nerlich and Jaspal, "Metaphors We Die By," 138.
45. Ibid., 139.
46. Ibid., 143.
47. See, for example, Margaret E. Mohrmann, *Medicine as Ministry: Reflections on Suffering, Ethics, and Hope* (Cleveland, OH: Pilgrim Press, 1995).
48. Andy Lipkis, "Afterword to the Twentieth Anniversary Edition," in Jean Giono, *The Man Who Planted Trees* (Burlington, VT: Chelsea Green, 2007), 49.
49. Drew Leder, ed., *The Body in Medical Thought and Practice* (Dordrecht: Springer, 2013), 3.
50. Ibid., 24.
51. Keekok Lee, *The Philosophical Foundations of Modern Medicine* (New York: Palgrave MacMillan, 2012), 25, 22.
52. Ibid., 63.
53. Ibid., 175.
54. Ibid., 176.
55. Garry Egger, "Dousing Our Inflammatory Environment(s): Is Personal Carbon Trading an Option for Reducing Obesity—and Climate Change?" *Obesity Reviews* 9 (2008).
56. Alan C. Moses, "Diabetes Prevention: Perspectives and Actions of One Company," *American Journal of Preventive Medicine* 44, 4 (2013).
57. The lifestyle alterations in question are extraordinarily difficult in a culture that does not encourage such healthy habits. I am not intending to blame diabetic patients for their disease. More on this, and on fat shaming, below.
58. Egger, 457.
59. Ibid., 458.
60. Ibid., 458–459.
61. Pamela D. Couture, "Partners in Healing: Bridging Pastoral Care and Public Health through Practical and Pastoral Theology," *Journal of Pastoral Theology* 5 (1995): 69.
62. See Steve De Gruchy, "Re-Learning Our Mother Tongue? Theology in Dialogue with Public Health," *Religion and Theology* 14 (2007); and Brenda Appleby and Nuala P. Kenny, "Relational Personhood, Social Justice, and the Common Good: Catholic Contributions toward a Public Health Ethics," *Christian Bioethics* 16, 3 (2010).
63. Risk factors have complex causes: the "food environment" (e.g., presence of grocery stores, food advertising, etc.) may relate to the prevalence of obesity, and the "built environment" (e.g., presence of sidewalks, parks, etc.) may relate to sedentary or active patterns of movement. (See K. Giskes et al., "A Systematic Review of Environmental Factors and Obesogenic Dietary Intakes among Adults: Are We Getting Closer to Understanding Obesogenic Environments?" *Obesity Reviews* 12, 5 [2011]. Also, C. P. Durand et al., "A Systematic Review of Built Environment Factors Related to Physical Activity and Obesity Risk: Implications for Smart Growth Urban Planning," *Obesity Reviews* 12, 5 [2011].)
64. Christian ethicist Richard O. Randolph notes, "There is an emerging class of ethical challenges in which both human and environmental health are interrelated, suggesting the need

for greater collaboration between Christian medical and environmental ethics. . . . These new ethical challenges actually present a cluster of interrelated, interdependent, mutually influencing issues" ("Human Health and Environmental Health Are Interdependent: Removing an Unnatural Partition within Bioethics," *Journal of the Society of Christian Ethics* 29, 1 (2009): 164.)

65. Stephen M. Gardiner, "Geoengineering and Moral Schizophrenia: What Is the Question?" in Wil C.G. Burns and Andrew L. Strauss, eds., *Climate Change Geoengineering: Philosophical Perspectives, Legal Issues, and Governance Frameworks* (Cambridge: Cambridge University Press, 2013).

66. Donna Haraway's "cyborg" concept might imply that there is something productive about humanity's hybridization with machines. I wonder if she would welcome a "cyborg" Earth, a geoengineered planet which hosts human life due to human technological intervention? (Donna Haraway, "A Cyborg Manifesto: Science, Technology, and Socialist-Feminism in the Late Twentieth Century," *Simians, Cyborgs, and Women: The Reinvention of Nature* [New York: Routledge, 1991].)

67. Gardiner, "Geoengineering and Moral Schizophrenia," 18.

68. Things might be different if, in a world of scarce resources, one could support *either* preventive measures *or* the development of dialysis. But in this world, there is space for both prevention and worst-case scenario research—in terms of diabetes and in terms of climate change.

I should note that geoengineering research as a process comes with certain risks and costs—where and how exactly do we test it?, for example—and those considerations are not in the realm of this paper, but are certainly worth weighing carefully before fully supporting geoengineering research.

69. Mike Hulme, *Can Science Fix Climate Change? A Case against Climate Engineering* (Cambridge: Polity Press, 2014), xii. Emphasis in original.

70. Keulartz, "Using Metaphors in Restoring Nature," 43.

71. Clive Hamilton, *Earthmasters. The Dawn of the Age of Climate Engineering* (New Haven: Yale University Press, 2013).

72. Merchant, *The Death of Nature*, 99–100.

BIBLIOGRAPHY

Appleby, Brenda, and Nuala P. Kenny. "Relational Personhood, Social Justice, and the Common Good: Catholic Contributions toward a Public Health Ethics." *Christian Bioethics* 16, 3 (2010): 296–313.

Boudry, Maarten, and Massimo Pigliucci. "The Mismeasure of Machine: Synthetic Biology and the Trouble with Engineering Metaphors." *Studies in History and Philosophy of Biological and Biomedical Sciences* 44 (2013): 660–668.

Clingerman, Forrest. "Seeking the Depth of Nature in a Scientific World." In *Technology, Trust, and Religion: Roles of Religions in Controversies on Ecology and the Modification of Life*, edited by William Drees, 141–156. Leiden University Press, 2009.

Couture, Pamela D. "Partners in Healing: Bridging Pastoral Care and Public Health through Practical and Pastoral Theology." *Journal of Pastoral Theology* 5 (1995): 65–80.

Durand, C. P., et al. "A Systematic Review of Built Environment Factors Related to Physical Activity and Obesity Risk: Implications for Smart Growth Urban Planning." *Obesity Reviews* 12, 5 (2011): 173–182.

De Gruchy, Steve. "Re-Learning Our Mother Tongue? Theology in Dialogue with Public Health." *Religion and Theology* 14 (2007): 47–67.

Egger, Garry. "Dousing Our Inflammatory Environment(s): Is Personal Carbon Trading an Option for Reducing Obesity—and Climate Change?" *Obesity Reviews* 9 (2008): 456–463.

Gardiner, Stephen M. "Geoengineering and Moral Schizophrenia: What Is the Question?" In *Climate Change Geoengineering: Philosophical Perspectives, Legal Issues, and Governance Frameworks*, edited by Wil C.G. Burns and Andrew L. Strauss, 11–38. Cambridge: Cambridge University Press, 2013.

Giskes, K., et al. "A Systematic Review of Environmental Factors and Obesogenic Dietary Intakes among Adults: Are We Getting Closer to Understanding Obesogenic Environments?" *Obesity Reviews* 12, 5 (2011): 95–106.

Hamilton, Clive. *Earthmasters: The Dawn of the Age of Climate Engineering*. New Haven: Yale University Press, 2013.

Haraway, Donna. "A Cyborg Manifesto: Science, Technology, and Socialist-Feminism in the Late Twentieth Century." *Simians, Cyborgs, and Women: The Reinvention of Nature*. New York: Routledge, 1991.

Hulme, Mike. *Can Science Fix Climate Change? A Case against Climate Engineering*. Cambridge: Polity Press, 2014.

Keekok Lee. *The Philosophical Foundations of Modern Medicine*. New York: Palgrave MacMillan, 2012.

Keulartz, Jozef. "Using Metaphors in Restoring Nature." *Nature and Culture* 2, 1 (Spring 2007): 27–48.

Kintisch, Eli. *Hack the Planet: Science's Best Hope—or Worst Nightmare—for Averting Climate Catastrophe*. New York: Wiley, 2010.

Lakoff, George, and Mark Johnson, *Metaphors We Live By*. University of Chicago Press, 1980.

Leder, Drew, ed. *The Body in Medical Thought and Practice*. Dordrecht: Springer, 2013.

Lovelock, James. *The Revenge of Gaia: Earth's Climate Crisis and the Fate of Humanity*. New York: Basic Books, 2006.

McFague, Sallie. *Models of God: Theology for an Ecological, Nuclear Age*. Minneapolis: Fortress Press, 1987.

———. *The Body of God: An Ecological Theology*. Minneapolis: Augsburg Fortress, 1993.

Merchant, Carolyn. *The Death of Nature: Women, Ecology, and the Scientific Revolution*. New York: Harper and Row, 1980.

Mohrmann, Margaret E. *Medicine as Ministry: Reflections on Suffering, Ethics, and Hope*. Cleveland, OH: Pilgrim Press, 1995.

Moses, Alan C. "Diabetes Prevention: Perspectives and Actions of One Company." *American Journal of Preventive Medicine* 44, 4 (2013): 333–338.

Myers, Teresa A., Matthew C. Nisbet, Edward W. Maibach, and Anthony A. Leiserowitz. "A Public Health Frame Arouses Hopeful Emotions about Climate Change." *Climatic Change* 113 (2012): 1105–1112.

Nerlich, Brigitte, and Rusi Jaspal. "Metaphors We Die By? Geoengineering, Metaphors, and the Argument from Catastrophe." *Metaphor and Symbol* 27 (2012), 131–147.

Nicholson, Daniel J. "Organisms ≠ Machines." *Studies in History and Philosophy of Biological and Biomedical Sciences* 44 (2013): 669–678.

Nisbet, Matthew C. "Communicating Climate Change: Why Frames Matter for Public Engagement." *Environment* 51, 2 (March 1, 2009): 14–23.

O'Brien, Kevin J. "The 'War' against Climate Change and Christian Eco-Justice: Ethical Implications of Martial Rhetoric." *Worldviews* 16 (2012): 135–153.

Porter, Kate Elizabeth, and Mike Hulme. "The Emergence of the Geoengineering Debate in the UK Print Media: A Frame Analysis." *The Geographical Journal* 179, 4 (December 2013): 342–355.

Randolph, Richard O. "Human Health and Environmental Health Are Interdependent: Removing an Unnatural Partition within Bioethics." *Journal of the Society of Christian Ethics* 29, 1 (2009): 153–170.

Romaine, Suzanne. "War and Peace in the Global Greenhouse: Metaphors We Die By." *Metaphor and Symbolic Activity* 11, 3 (1996): 175–194.

Chapter Eight

Stewards of Irony: Planetary Stewardship, Climate Engineering, and Religious Ethics

Willis Jenkins

If humans were to cut down the world's forests and convert as much land as possible to biofuel production, would that counteract dangerous dynamics of climate change? I am part of a multidisciplinary research team working on that question at the University of Virginia (UVa). We run global climate models to investigate how large-scale changes in management of the earth's surface would affect climate dynamics.[1] Our team wants to understand how various allocations of earth's surface to food agriculture, biofuel production, and ecological reserves correlate with climate outcomes. Basically, we are running geoengineering simulations in which land surface management is the instrument of intentional influence.

Our project should raise disturbing questions. Wouldn't climate forcing by total manipulation of earth's surface display the worst vices of geoengineering? Responding to excessive human interference in one global system by intensifying interference in another seems to perversely double down on human domination. Our simulations suppose that every square kilometer of land surface is drafted into service as a tool for managing climate. Moreover, they require a totalitarian conceit; positing ideal allocations of land use blithely skips questions of governance and culture that would make implementation of any scenario dystopically unjust. As with geoengineering generally, the question must be asked: does it not corrupt our moral and political imaginations even to entertain such possibilities?

A question perhaps even more disturbing to some: what is a religious ethicist doing on a team that runs climate models? Beyond pronouncing the outcome of a model simulation as "bad" or "good" with all the authority vested in a professional ethicist, does my training make any real contribution to the knowledges required to build and run climate models? Worse than irrelevance may be complicity: perhaps the participation of a scholar of religious ethics, even if otherwise useless, confers a kind of moral legitimacy on geoengineering projects?

My answer to those two sets of questions is connected: eliciting properly disturbed responses to geoengineering research may be helped by an interpretive scope often (although not necessarily or exclusively) cultivated by religionists. This chapter focuses on developing criteria of judgment through the interpretive irony and cultural play afforded by attention to narratives about humanity and earth in evolutionary history. Those "big stories," I argue, do not automatically authorize or prohibit any particular geoengineering proposal; supposing that they can is in fact a common mistake made in religious ethics. Attending to how stories narrate the integrity of human action on a planetary scale can, however, help various publics consider what criteria they have (or need) to evaluate research on climate engineering and other forms of planetary management.

As with research on other boundary technologies (like genetic engineering), research on geoengineering should help cultures develop criteria by which various schemes could be evaluated. In conditions of ambiguity, uncertainty, and rapid change, societies may not have the concepts, let alone the consensus, on how to interpret new responsibilities. So research on such technologies needs to help cultures learn how to interpret them. Clearly we need public debate about thresholds of acceptability for implementing various climate engineering technologies. Who gets to decide and by what criteria?[2] Developing those criteria may happen by bringing moral uncertainty into the research process itself.

Nancy Tuana has made especially compelling arguments for integrating ethical analysis into geoengineering research teams, showing how involvement of ethically trained humanists can help produce better knowledge. For example, where a research team might think it better to avoid drawing attention to background judgments that shaped the bounds of its inquiry or to ambiguous decisions it had to make about interpreting and presenting data, an ethicist might help the team develop prose accounts of epistemic, moral, and political uncertainties in ways that deepen civic understanding.[3] Described well, uncertainties can stimulate cultural learning about the context of the research. On my own team, it seems impossible to integrate into a computer model the factors that shape how people actually make land use decisions across the planet. Drawing attention to that limit could help publics think beyond specific outcomes of our models to imagine how current land uses already function as a form of climate cultivation.

Interpreting the context of questions about climate engineering is a major aspect of developing criteria for judging various proposals. The metaphor of "cultivation" shapes a certain way of thinking about the human relationship with the atmosphere. We cultivate land (agriculture) and sea (aquaculture), argues Mike Hulme, and now we have begun to recognize that humans cultivate the sky ("weatherculture"). As with agriculture, then, the question is not *if* humans should be involved, but what the criteria are for good involvements. Hulme asks, "What sort of cultivational aspirations and practices is humanity on the verge of implementing?"[4] The sort of criteria we need to make sense of climate engineering, Hulme suggests, have less to do with climate outcomes than with the kind of human-climate relationship we want. The UVa modeling exercise can help us think about the connection between aspirations for two relationships: what we do with land and what we do with climate.

The metaphor of cultivation assumes that humanity is already involved in producing weather, that human action is already an aspect of climate (the cultural expectation of weather.)[5] So the metaphor carries the beginning of a story about the changing human relationship with earth. As Laura Hartman explores in this volume, different metaphors give rise to different criteria for evaluating climate engineering because they carry different pictures or story lines about the human relation with earth. Those metaphors drive the cultural frames that shape how questions come to a public. "Frames are metaphors writ large," writes Hartman, following the definition of Mathew Nisbet, who calls frames "interpretive storylines that set a specific train of thought in motion."[6]

Using a metaphor of cultivation to shift the question from *whether* humans should manage the climate to *how* to do so well in fact follows a train of thought established by advocates for geoengineering. Advocacy arguments, show Jonas Anshelm and Anders Hansson, usually follow one of several major narratives, all of which rely on the basic notion that humans are already influencing climate and could at least try to do so more responsibly. For example, here is a story line that Anshelm and Hannson call "double fear:" once upon a time most scientists thought geoengineering a bizarre notion, but now, made desperate by looming catastrophe, they are driven to geoengineering as an emergency measure.[7]

Is that the right story line about the context of climate engineering questions? One problem with the emergency story and the cultivation metaphor is that they seem to elide the unease with which many people regard the situation in the first place. How did we get ourselves into a situation in which it makes sense for researchers to investigate ways of manipulating planetary systems? Cultural unease at our predicament should at least inflect criteria for evaluating geoengineering proposals. In fact, the most significant knowledge produced by some forms of geoengineering research may be the disturbances of cultural cognition they set off.

No one on the UVa team thinks that the scenarios we investigate are politically sane or morally justified. To fiddle with the earth's surface in a computer simulation does not mean to propose actually managing every square kilometer of the globe in a way to solve the climate crisis. What kind of people would do that? We want the simulations to inform understandings of global interactions between climate policy goals and land uses, and at a deeper level to usefully disturb cultural cognition about human use of the planet. Yet some publics might seize on the research to suggest the availability of land-as-engineering solutions. (All that we lack is control over all the land on the planet.) So how can the team conduct its research in a way that illuminates understanding while also inviting criticism, if not darkly ironic suspicion, of the sort of story about humanity into which direct application of our research would fit?

CORRUPTION, IRONY, AND RELIGION

Many critics point to the moral hazard of researching geoengineering at all. Does geoengineering corrupt the political will and moral imagination needed to undertake more responsible and still available measures? In response to researchers who argue that preparing for emergency by developing the least-worst options is merely prudent, Stephen Gardiner observes that "there seems to be an important moral difference between (on the one hand) preparing for an emergency and (on the other hand) preparing for an emergency that is to be brought about by one's own moral failure."[8] If its premise involves a calculated failure of responsibility, then geoengineering research must—at least, and additional to whatever else it does—find ways to interrogate the conditions that make its questions plausible. The condition of calculated moral failure in which the question arises makes Gardiner ask about the character of humanity: "What kind of people would make the choice to geoengineer?"[9]

The fecundity of that question may warrant its risk of corruption. If researchers foreground the moral hazard of their research, its risks could function as a stimulus to more adequate deliberation. In the case of climate change, many of the practical decisions involve interpreting the big background story of humanity and earth by which they make sense. As Anshelm and Hannson close their essay on story lines, they expand the narrative scale entailed in making sense of them: "discussing geoengineering inevitably entails negotiation of what it is to be human on a truly grand scale."[10] If that is so, then developing criteria of evaluation for geoengineering research needs to elicit public engagement with basic questions about the human condition. How can it do that?

A clue lies in the irony with which the UVa researchers regard our model's assumptions. Perhaps we should present the models in ways that invite people to consider the big story in which the research scenarios *would* make sense. What sort of story about humanity and earth would become true if they were in fact implemented? Questions of such interpretive scope are usually excluded from research reports, but in the case of geoengineering (as Bruhn et al. show in this volume), a transdisciplinary approach can facilitate productive exchange with civil society.[11] Considering narratives about humanity and earth that could make sense of geoengineering can help draw diverse public engagement with research that otherwise may seem daunting precisely because of its overwhelming scale or odd assumptions. Big stories can help interpret the planetary scope and evolutionary scales involved in geoengineering projects and begin to develop evaluative criteria for making sense of them.

Religionists are supposed to be good at facilitating that sort of attention.[12] Particular religious stories can also help stimulate that interpretive scope, even and perhaps especially when people regard them with ironic distance. Consider Kevin O'Brien's presentation of the context of climate engineering in terms of a theological story about violence and sin in which appropriate action begins with repentance and aims for restoration.[13] One need not identify with Christianity in order to appreciate how this interpretive framing integrates critical suspicion of the context while also imaging the kind of people (repentant) who could conceivably deploy climate engineering with integrity. Insofar as that story captures something that seems true of the social and ecological relations involved in geoengineering, it may help people negotiate how to be human on a planetary scale, even if they do not hold all the beliefs or commitments belonging to that particular metanarrative. Something similar holds for the *Pachamama* cosmologies with which some indigenous peoples frame climate questions: they interpret the context of climate engineering in a way that may help people think through the social and ecological relations involved in becoming creatures who do that kind of thing.

Because they involve views about the past and future of the human story in earth, debates over geoengineering involve debates over moral cosmology. At stake are imaginative construals of humanity's embodiment in the world and narratives that lend meaning and purpose to how we live out that existence. Without supposing that such construals necessarily arise from conventionally religious sources (I soon turn to evolution as a possible source), geoengineering research cannot avoid questions with a religious heft. Climate change is "a question of ultimacy," writes Forrest Clingerman, in that it is a site of conflicting interpretations about what it means to be human. As climate compels reinterpretations of the human relation to earth, Clingerman suggests, religionists can help mediate conflicting narratives.[14] My role on the research team might then be to make the moral hazard of geoengineering research elicit properly scaled conversations about cosmology.

Now, religionists love to transform practical problems into religious problems by insisting that their roots lie in metanarratives. I have in fact argued that religionists should beware of that "cosmological temptation," in part because framing issues in terms of conflicting worldviews can turn ameliorable problems in need of many knowledges into intractable metaphysical conflicts.[15] Adding religious conflict to climate engineering debates is certainly no guarantee of improving them.

However, my experience with the climate modeling team has underscored for me a practical function of big stories: they afford researchers conceptual space in which to spell out possible implications of research assumptions or to imaginatively inhabit the basic relations in some alternative story. Moreover, thinking with big stories can stimulate interpretive play through which participants test rival sets of criteria, imagining what they imply about the kind of humans who would use them. A shortsighted version of pragmatism would suppress interpretive play by bracketing cosmology in order to deploy reductive tools of decision-making—like ranking climate engineering schemes by cost-benefit analysis.[16] Moral hazard runs highest precisely when cosmology is suppressed, for then metaphors and frames escape scrutiny, and the liability of the entire question to conflicting interpretations of the human condition is obscured. What kind of people would manipulate the entire surface of earth in order to control the planet's climate? One answer: people who avoid questions like that.

The question should elicit ironic disturbance. Geoengineering represents an effort to take responsibility for cultivation of the sky, but the fact we need to do so should make us question whether we remain competent to act responsibly. Do we (and here I mean citizens of cultures who have destabilized global environments) really know what would it mean to act responsibly? Jonathan Lear's *Case for Irony* describes the experience of being grabbed by a sense that one's entire range of practical striving falls short of the ideals embedded in it. His example: a very good teacher by all conventional measures may suddenly find herself wondering what it really means to teach. "The possibility of irony arises when a gap opens between pretense as it is made available in a social practice and an aspiration or ideal which, on the one hand, is embedded in the pretense . . . but which, on the other hand, seems to transcend the life and the social practice in which that pretense is made."[17] Geoengineering is a more compromised situation than Lear's teacher represents, but this aspect holds similar: geoengineering research is a social practice of responsibility in which participants might be led to question what it really means to act responsibly. What greater pretense could there be than for a creature so disruptive of earth's ecology to present itself as steward of the planet? By drawing out stories that can make sense of such tensions, religionists may be able to facilitate the sort of irony that forestalls the moral and political corruption to which such research is liable.

PLANETARY STEWARDSHIP?

Irony of that sort seems missing from the geoengineering research that has so far had the greatest cultural impact. The scientist most responsible for making geoengineering a serious proposal is also the figure most responsible for introducing the Anthropocene as the story line by which to make sense of it. Paul Crutzen's 2006 article in *Climatic Change*, "Albedo Enhancement by Stratospheric Sulfur Injections," made a technological proposal that seems fitting by light of the historiographic proposal made in his 2002 note in *Nature*, "Geology of Mankind."[18] The first article argued for adopting a planetary perspective characterized by human domination, then the 2006 essay proposed a technology for coping with unintended consequences of human influence through intentional application of planet-scale technology. Geoengineering and the Anthropocene were made for each other; the two articles have been cited thousands of times, the idea in one driving a sense of inevitability in the other.

The idea of the Anthropocene seems like an astonishing gift to the humanities, for with it a scientist puts the human back in the center of scientific research—and thus also gives new legitimacy to the interpretive knowledges of the humanities. No one has received the gift more enthusiastically than the historian Dipesh Chakrabarty, who reports that the Anthropocene thought overturned basic assumptions informing his work and redirected his focus. (It grabbed him in the way Lear describes, making him wonder what it really means to do history.) For history now finds itself within a bigger story in which objects of the natural sciences become human products (anthropogenic climate change) and the subject of the humanities has become a geological agent. In fact, says Chakrabarty, climate change only appears as a problem when the frame of human history is widened by geological perspective: "Without such knowledge of the deep history of humanity it would be difficult to arrive at a secular understanding of why climate change constitutes a crisis for humans." By "secular," Chakrabarty seems to intend "non-religious," but his thought involves a shift that looks religious in register: "The consequences only make sense if we think of humans as a form of life and look on human history as part of the history of life on this planet."[19] The cosmological reorientation required to achieve his secular understanding seems religious in depth, in the sense that it involves a big story that shapes interpretation of the human relationship and construes the context of human action.

The idea of the Anthropocene can carry a stealth religiosity—an imaginative construal of the human place and purpose with the planet that goes unrecognized as a particular moral cosmology. When presented as a natural fact about the world, its interpretive choices and imaginative consequences may remain concealed, suppressing the sort of ironic play needed to reflect on the way it frames the human condition. Within the Anthropocene imagination, humanity has a story as a species because it now acts as a geological agent, which it discovers through the event of climate crisis. In the context of that crisis, geoengineering seems merely prudent, responsible, and inevitable. (Crutzen's retort to those who think we should still focus on drawing down greenhouse gas emissions—"looks like a pious wish"—suggests nostalgia for an old faith.)[20] Yet, as Clingerman illuminates in this volume, geoengineering enacts the new temporal sensibility in which its practice could appear responsible; it represents a domestication of time that goes with Crutzen's sense that the Anthropocene should become a story of wise control over earth.[21]

The narrative background must not go unquestioned. If the species is an agent in a story, then it makes sense to ask a character question: into what sort of story about humanity does this practice fit? That question makes possible an ironic sense about the fragility of our interpretations of ourselves and our world, which can make us more accountable for the imaginative construals carried by our practices. For that, however, we must be explicit about the pretense to responsibility in our practices of global environmental management, and the practical ideals by which they are oriented.

Consider how the virtue trope of stewardship organizes a major research frame for environmental science. Climate change, write several prestigious teams of scientists, requires rethinking humanity's role in planetary systems.[22] Since we are cultivating the planet one way or another, they argue, science should lead efforts to actively transform ecological systems and thereby reshape the vision of human life within them. Humans must exercise "planetary stewardship."

In a bit of circular reasoning, taking responsibility for anthropogenic influence over the planet requires developing stewardship at the level of the species, yet the virtue is indexed to human well-being. Planetary stewardship, they say, shifts the focus of science from supplying information to manage problems in reference to historic benchmarks toward "the active shaping of trajectories of change in coupled social-ecological systems . . . to enhance ecosystem resilience and promote human well-being."[23] What nature would be without human interference no longer matters; human flourishing now normatively regulates what an ecosystem should do. The virtues of ecological care are indexed to a broader view of the virtues of being human. What ecological conditions should humanity seek? The sort "within which humanity can continue to develop in a humane and respectful fashion."[24] So is geoengineering humane and respectful? That depends on what we make of our pretense to acting "humanely" in the Anthropocene.[25]

Rather than trying to settle the circularity with a definition, I think, it is better to ask: what sort of practices are most apt for helping cultures think through that interpretive question? At least, they cannot be practices in which it seems the pretense is secured from critical wonder. Whatever course humanity takes, we must be able to keep wondering: what does it really mean to act as steward of the planet?

That sort of irony may be facilitated by conflict at the level of religious cosmology. Consider Eileen Crist's trenchant criticism of Anthropocene discourse. She worries that it legitimates continued expansion of human power over earth and, more insidiously, ends argument between ecocentrism and anthropocentrism by making the latter appear a natural fact (if human interests rule the world in fact, there is no longer a question about whether they should). Crist protests the way that the Anthropocene story line makes climate engineering appear inevitable and alternatives a pious wish.[26]

A different narrative would make those alternatives seem more attractive and feasible. Crist points attention to Thomas Berry's proposal to call this epoch the "Ecozoic," which better fits his sense of humanity place in an integrated community of life (famously, "a communion of subjects"). Crist thus prefers a story that imagines humanity's smallness within the community of life, rather than one that emphasizes human immensity. Instead of leaning into "active stewardship" on an Anthropocene narrative, Crist wants a story in which it makes more sense for humanity to withdraw from vainly attempting to govern earth systems. Humans should learn that responsibility means deferring to the long evolutionary history of which we are a part.

Ecological modesty may then require narrative audacity. Berry understood that embracing the epistemic and moral humility entailed in his sense of ecological communion would require a very big story. Berry compared his project, "The Great Work," to that of Augustine, whose *City of God* resets the narrative context in which humane civilization could recognize itself after the fall of Rome.[27] Crist is less ambitious but seems to need a similar theological register for the summation of her argument: "In making ourselves integral, and opening into our deepest gift of safeguarding the breath of Life, the divine spirit of the human surfaces into the light."[28] What is the divine spirit of the human? She does not need to be totally clear about that for the reader to understand that it involves the sort of big story in which geoengineering appears repulsive. Her religious mode of engagements forces the question: would we not betray what it means to be human?

At stake in geoengineering, then, is not merely stewardship of the planet but stewardship of the idea of humanity by which acts of planetary responsibility will be interpreted. Detractors and proponents of geoengineering agree that humanity has been indifferent to the climatic consequences of its behavior and must now take responsibility. They differ in crucial ways over what it means to be responsible because they differ in how to interpret humanity and its relation to its environment. Whether and how humans geoengineer the planet will be a major factor in determining not only the future of life's evolution on Earth, but the future of humanity's understanding on itself.

POSTRELIGIOUS COSMOLOGY

Is there a way to engage the cosmological dimensions of geoengineering questions that avoid typical cultural conflicts over religion? In modern cultures, "religion" often stands for rivalrous views of the good that, because their conflict seems undecidable, should stay out of public discussion. How can geoengineering research elicit discussion about the human condition without introducing intractable conflict? One major option for attempts to develop a big story from evolutionary cosmology, looking to science for postreligious foundation. I briefly illustrate that option and then argue for a second, more pluralist approach in which particularist constructions open multiple possibilities of cultural learning through ironic play.

Ursula Goodenough's *Sacred Depths of Nature* displays the first option elegantly. Humanity possesses the foundations for a planetary ethic, she says: "The story of Nature has the potential to serve as the cosmos for the global ethos that we need." The narrative context needed to make sense of planetary responsibility may be found in the epic of evolution. Precisely because diverse cultures now share responsibility for one atmosphere, we need a common cosmology; one shared story for one shared world. The function of religion is to make that evolutionary epic come alive within individual persons; "the role of religion is to integrate the Cosmology and the Morality, to render the cosmological narrative so rich and compelling that it elicits our allegiance and commitment to its emergent moral understandings."[29] Brian Swimme and Mary Evelyn Tucker in *Journey of the Universe* fulfill that affective role by presenting evolutionary cosmology as an unfolding story that can orient right and fitting forms of human creativity.[30] Because the story belongs to everyone, it holds all the other big stories of religions and cultures, which can each find their way to fit in it. It is "postreligious," in that sense.

So then, should humans exercise that creativity by terraforming the planet to manage the climate? The story underdetermines ethical choice. One could make the case that manipulating the planet's atmosphere represents hubris and vain grasping for control, and would therefore be an unfitting participation in the story of the universe. But one could as well make the case that geoengineering represents the Earth self-consciously directing its own future.

Both are plausible implications of the narrative; neither is required nor ruled out. Lisa Sideris in fact argues that looking to evolution for a single guiding story might in fact undermine public ethical deliberation, since so much energy is spent achieving an abstract consensus that still underdetermines any practical choice.[31]

Consider David Christian's *Maps of Time,* which he describes as "big history." It is, he says, "a modern creation myth," speaking to the human need for belonging; "coordinates within which people can imagine their own existence and find a role."[32] The promise of a big history is that it can help humanity see itself as a whole and reckon with the great uncontrolled experiment it is undertaking with earth in this "utterly new stage in human history and in the history of human relations with other species and with the earth . . . a decisive moment in the history of the entire biosphere."[33] Again, however, the big story puts the geoengineering question in context only to overwhelm the question with a scale that could let the current human moment open in many different directions. Geoengineering seems to fit with that history as easily as refusing to do so. What is the use of constructing a global myth for such an indeterminate ethic?

Maybe we do not need consensus on one big story because the relevant sort of religiosity here is not about finding foundations for ethics, but is rather about a kind of play with possible human fates. A second approach supposes that, instead of asking whether a story justifies terraforming the planet with biofuels, we could rather ask: what stories would be told about the humans who undertook such geoengineering? Dialogue about those stories can help people of many cultural worlds think together about how the geoengineering practices interpret the human condition, which may be sufficient for retaining accountability for the cosmological implications, even without consensus.

There is evolutionary warrant for treating the epic of evolution as one candidate narrative for eliciting appropriate interpretive play. "When it comes to telling big stories about the order of existence," writes Robert Bellah, "then, even if they are scientific stories, they will have religious implications."[34] Bellah's *Religion in Human Evolution* follows evolutionary theories that explain how play allows organisms to create an environment that supports new developmental and evolutionary directions. Play seems to create a relaxed field of competition which can support the development of capacities that make organisms and perhaps species more successful. Bellah then suggests that ritual might be thought of as a kind of cultural play, allowing symbol-making creatures a relaxed field in which to pursue new possibilities.

Religion might then be considered as a kind of adaptive nonfunctionalism by supporting ventures of symbolic transcendence that allow humans to interpret the problems of their existence in new ways. Religion on that view affords irony, in the sense I have used it in this chapter. New biocultural possibilities for civilizations may be opened or at least made intelligible by symbolic play that, in effect, uses locally approved material (like evolutionary science or Catholicism) to rethink everything about a culture, including its way of imagining the human condition.

Bellah closes his deep history of religion by explaining what this perspective can do for humanity's planetary challenges and irreducible pluralism. Deep history puts in perspective humanity's sudden destruction of earth's biodiversity, he writes, and it allows its students to accept religious pluralism. Many human traditions of meaning, each explicable on its own terms, are just what one would expect from a long and open field of symbolic play. Deep history lets scholars give up the search for one absolutely true myth (postreligious or not) by which to measure the story of humanity. And it lends no support for thinking that we can do without myth—that we have a mythless backup system to settle decisions by consequentialist prudence.[35]

We should not expect humanity to ever have a single big story; moreover, if we did, that story could not settle questions like geoengineering. We may be a geological species with a universe story, but we are always also cultural persons living in a moral pluriverse. We make sense of the problems of the species through the many distributed forms of play afforded by the pluriverse. Good play, adaptive play, should regularly elicit ironic precarity; it should make us wonder, do we know what it would mean to steward the planet or engineer the climate?

As one cosmological cluster in the human pluriverse, consider the kind of stories that the Abrahamic traditions could tell of humans who geoengineer. These traditions can ask: how would God make sense of humans terraforming earth in order to manage the climate? The question hardly settles anything—maybe God would delight in it, maybe God would abhor it, or maybe God would shrug. But the question does shift the framing in a way that refuses to let all the agency reconcentrate in human hands. By imagining how God would tell the story, interpretive criteria for practices of climate responsibility refer to something more than the human measure—to one who rather takes the measure of humanity. One need not be a member of one of those traditions, or even a believer in God, to stimulate ironic reflection on what it really means to act responsibly.

Again, Abrahamic play has no special authority for initiating that kind of culture-questioning ironic transcendence. Indigenous *Pachamama* cosmologies will interpret the context differently, as will the epic of evolution. It might work to ask what story we want future generations to tell of us, how they should interpret our sense of responsibility and its failures. The point is that eliciting interpretive play at the level of big stories can help geoengineering research drive social discussion while limiting its moral hazard. Recall Gardiner's concern that geoengineering research may permit calculated moral failure by letting us plan for the consequences of failing to take responsible action. A sense of ironic precarity about our concepts of what it means to act responsibly can help forestall that hazard.

CONCLUSION

So, should humans geoengineer Earth's atmosphere by terraforming its surface? I have argued that the absurdity of our research questions should shock persons into something approaching religious reflection. Maybe it helps to have a religionist around to invite that kind of ironic play. Responsibility for climate change begins in recognition that humans are already remodeling Earth's atmosphere, and doing so without much character—without a sense of the story in which doing so would be intelligible.

By illuminating climatic implications of global land use decisions, the modeling research at UVa underscores the way local decisions to plow a field or plant a forest cumulatively shape the impact of a geological agent. Simulations of massively different land allocations should not feed totalitarian political imaginations; they should question the sort of predicament that leads to such questions. What sort of people find themselves in such a mess? In what sort of story does a problem like that appear?

Aldo Leopold's closing line in *Sand County Almanac* laments that "we are remodeling the Alhambra with a steam shovel and we are proud of our yardage."[36] Steam shovels might sometimes be used, Leopold grants, but not by the sort of people so disoriented from relevant criteria for remodeling sacred geography that they are proud of *yardage*. Maybe geoengineering might be something people use, but not with such obtuse criteria for how to deploy it well. We may have to use steam shovels to respond well to climate change, but we won't know whether we are responding well or not by measuring relative reductions in radiative forcing or calculating least-worst options of response. That is just yardage. Criteria for a decent proposal for intentionally altering Earth's atmosphere include accountability to the commitments and practices by which humanity sustains the myths of our own humaneness. What kind of people would intentionally cultivate the sky? The most tragic, most corrupting course of action would lie in avoiding such questions.

NOTES

1. Directed by Deborah Lawrence; for background see Deborah Lawrence and Karen Vandecar, "Effects of Tropical Deforestation on Climate and Agriculture," *Nature Climate Change* 5:1 (2014).

2. Arguments for and against different kinds of proposals already make a substantial literature; for representative overview see David Keith, *A Case for Climate Engineering* (MIT,

2013); Philip Boyd, "Ranking Geo-Engineering Schemes," *Nature Geoscience*, 1 (2008); Christopher Preston, "Ethics and Geoengineering: Reviewing the Moral Issues Raised by Solar Radiation Management and Carbon Dioxide Removal," *Climate Change* 4 (2013).

3. Nancy Tuana et al. "Towards Integrated Ethical and Scientific Analysis of Geoengineering," *Ethics, Policy, & Environment* 15, 2 (2012).

4. Mike Hulme, "Better Weather? The Cultivation of the Sky?" *Cultural Anthropology*, 30, 2 (2015): 241.

5. Ibid., 238.

6. Laura Hartman, "Healing the Climate," chapter 7 in this volume.

7. Anders Hansson and Jonas Anshelm, "The Last Chance to Save the Planet? An Analysis of Geoengineering Advocacy," *Environmental Humanities* 5 (2014).

8. Stephen Gardiner, "Is Arming the Future with Geoengineering Really the Lesser Evil?," in *Climate Ethics: Essential Readings,* Stephen Gardiner, ed. (New York: Oxford, 2010), 293.

9. Ibid., 303.

10. Anshelm and Hannson, "Last Chance," 118.

11. Thomas Bruhn et al., "From the Garden of Eden to Eden's Gardener," chapter 2 in this volume.

12. See Matthew Nisbet et al., "Four Cultures: New Synergies for Engaging Climate Change," *Frontiers in Ecology and the Environment* 8 (2010): 329–31.

13. Kevin O'Brien, "First Be Reconciled," chapter 10 in this volume.

14. Forrest Clingerman, "Theologians as Interpreters—Not Prophets—in a Changing Climate," *Journal of the American Academy of Religion* 83, 2 (2015): 337.

15. Willis Jenkins, *The Future of Ethics: Sustainability, Social Justice, and Religious Creativity* (Washington DC: Georgetown University Press, 2013).

16. With appeal to findings from what brains do when presented with complex scenarios in imaging scanners, Joshua Greene argues that consequentialism functions back-up moral cognition system within persons and should so function for global societies facing common problems like climate change. Whatever the warrant for his view of moral cognition, turning to consequentialism to decide whether to geoengineer still only makes sense within a certain construal of the human predicament with climate change. Whether consequentialism or geoengineering seems fitting, in other words, still depends on how we imagine ourselves—including what we make the relation between how our brains have evolved and what responsibility we have for the climate. Joshua Greene, *Moral Tribes: Emotion, Reason, and the Gap between Us and Them* (New York: Penguin, 2013).

17. Jonathan Lear, *A Case for Irony* (Cambridge: Harvard Press, 2011), 11.

18. Paul Crutzen, "Albedo Enhancement by Stratospheric Sulfur Injections: A Contribution to Resolve a Policy Dilemma?" *Climatic Change* 77 (2006); Paul Crutzen, "Geology of Mankind," *Nature* 415, no. 6867 (January 3, 2002). Crutzen had published the idea earlier, but the *Nature* article set off debate. See Will Steffen, Jacques Grinevald, Paul Crutzen, and John McNeill, "The Anthropocene: Conceptual and Historical Perspectives," *Philosophical Transactions of the Royal Society of London A: Mathematical, Physical and Engineering Sciences* 369, no. 1938 (March 13, 2011).

19. Dipesh Chakrabarty, "The Climate of History: Four Theses," *Critical Inquiry* 35, no. 2 (January 1, 2009): 213.

20. Crutzen, "Albedo Enhancement."

21. Forrest Clingerman, "Time's Arrow," chapter 5 in this volume.

22. Will Steffen et al., "The Anthropocene: From Global Change to Planetary Stewardship," *Ambio* 40 (2011); Carl Folke et al., "Reconnecting to the Biosphere," *Ambio* 40 (2011); F. Stuart Chapin et al., "Earth Stewardship: A Strategy for Social-Ecological Transformation to Reverse Planetary Degradation," *Journal of Environmental Studies & Sciences* 1 (2011); Chapin et al., "Ecosystem Stewardship: Sustainability Strategies for a Rapidly Changing Planet," *Trends in Ecology and Evolution* 25, 4 (2009).

23. Chapin et al., "Earth Stewardship," 45.

24. Steffen et al., "The Anthropocene," 756.

25. See Gisli Palsson et al., "Reconceptualizing the 'Anthropos' in the Anthropocene: Integrating the Social Sciences and Humanities in Global Environmental Change Research," *Environmental Science & Policy* 28 (2013).
26. Eileen Crist, "On the Poverty of Our Nomenclature," *Environmental Humanities* 3 (2013).
27. Thomas Berry, *The Great Work: Our Way into the Future* (New York: Bell Tower, 1999).
28. Eileen Crist, "On the Poverty of Our Nomenclature," 144.
29. Ursula Goodenough, *The Sacred Depths of Nature* (New York: Oxford University Press, 1998), xvii.
30. Brian Swimme and Mary Evelyn Tucker, *Journey of the Universe* (New Haven: Yale University Press, 2011).
31. Lisa Sideris, "Science as Sacred Myth? Ecospirituality in the Anthropocene Age," in Richard Rozzi ed., *Linking Ecology and Ethics for a Changing World* (Dordrecht: Springer, 2013).
32. David Christian, *Maps of Time: An Introduction to Big History* (Berkeley: University of California Press, 2011), 2.
33. Ibid., 440–41.
34. Robert Bellah, *Religion in Human Evolution: From the Paleolithic to the Axial Age* (Cambridge, MA: Harvard, 2011), 45–46.
35. Ibid., 606.
36. Aldo Leopold, *Sand County Almanac* (New York: Oxford, 1949), 226.

BIBLIOGRAPHY

Bellah, Robert N. *Religion in Human Evolution: From the Paleolithic to the Axial Age*. Cambridge, MA: Harvard University Press, 2011.

Berry, Thomas. *The Great Work: Our Way into the Future*. New York: Bell Tower, 1999.

Boyd, Philip W. "Ranking Geo-Engineering Schemes." *Nature Geoscience* 1, no. 11 (2008): 722–24.

Chakrabarty, Dipesh. "The Climate of History: Four Theses." *Critical Inquiry* 35, no. 2 (2009): 197–222.

Chapin, F. Stuart, Stephen R. Carpenter, Gary P. Kofinas, Carl Folke, Nick Abel, William C. Clark, Per Olsson, D. Mark Stafford Smith, Brian Walker, Oran R. Young, Fikret Berkes, Reinette Biggs, J. Morgan Grove, Rosamond L. Naylor, Evelyn Pinkerton, Will Steffen, and Frederick J. Swanson. "Ecosystem Stewardship: Sustainability Strategies for a Rapidly Changing Planet." *Trends in Ecology & Evolution* 25, no. 4 (2010): 241–49.

Chapin, F. Stuart, Steward T. A. Pickett, Mary E. Power, Robert B. Jackson, David M. Carter, and Clifford Duke. "Earth Stewardship: A Strategy for Social-Ecological Transformation to Reverse Planetary Degradation." *Journal of Environmental Studies and Sciences* 1, no. 1 (2011): 44–53.

Christian, David. *Maps of Time: An Introduction to Big History*. Berkeley: University of California Press, 2004.

Clingerman, Forrest. "Theologians as Interpreters—Not Prophets—in a Changing Climate." *Journal of the American Academy of Religion* 83, no. 2 (2015): 336–55.

Crist, Eileen. "On the Poverty of Our Nomenclature." *Environmental Humanities* 3 (2013): 129–47.

Crutzen, Paul J. "Geology of Mankind." *Nature* 415 (2002): 23.

Crutzen, Paul J. "Albedo Enhancement by Stratospheric Sulfur Injections: A Contribution to Resolve a Policy Dilemma?" *Climatic Change* 77, no. 3–4 (2006): 211–20.

Folke, Carl, Åsa Jansson, Johan Rockström, Per Olsson, Stephen R. Carpenter, F. Stuart Chapin, Anne-Sophie Crépin, Gretchen Daily, Kjell Danell, Jonas Ebbesson, Thomas Elmqvist, Victor Galaz, Fredrik Moberg, Måns Nilsson, Henrik Österblom, Elinor Ostrom, Åsa Persson, Garry Peterson, Stephen Polasky, Will Steffen, Brian Walker, and Frances Westley. "Reconnecting to the Biosphere." *Ambio* 40, no. 7 (2011): 719–38.

Gardiner, Stephen Mark. *Climate Ethics: Essential Readings*. New York: Oxford University Press, 2010.

Goodenough, Ursula. *The Sacred Depths of Nature*. New York: Oxford University Press, 1998.

Greene, Joshua David. *Moral Tribes: Emotion, Reason, and the Gap between Us and Them*. New York: Penguin, 2013.

Hansson, Anders, and Jonas Anshelm. "The Last Chance to Save the Planet? An Analysis of Geoengineering Advocacy." *Environmental Humanities* 5 (2014): 101–23.

Hulme, Mike. "Better Weather? The Cultivation of the Sky." *Cultural Anthropology* 30, no. 2 (2015): 236–44.

Jenkins, Willis. *The Future of Ethics: Sustainability, Social Justice, and Religious Creativity*. Washington, DC: Georgetown University Press, 2013.

Keith, David W. *A Case for Climate Engineering*. Cambridge, MA: MIT Press, 2013.

Lawrence, Deborah, and Karen Vandecar. "Effects of Tropical Deforestation on Climate and Agriculture." *Nature Climate Change* 5, no. 1 (2014): 27–36.

Lear, Jonathan. *A Case for Irony*. Cambridge, MA: Harvard University Press, 2011.

Leopold, Aldo. *A Sand County Almanac, and Sketches Here and There*. New York: Oxford University Press, 1949.

Nisbet, Matthew C, Mark A Hixon, Kathleen Dean Moore, and Michael Nelson. "Four Cultures: New Synergies for Engaging Society on Climate Change." *Frontiers in Ecology and the Environment* 8, no. 6 (2010): 329–31.

Palsson, Gisli, Bronislaw Szersynski, Sverker Sörlin, John Marks, Bernard Avril, Carole Crumley, Heidi Hackmann, Paul Holm, John Ingram, Alan Kirman, Mercedes Pardo Buendía, and Rifka Weehuizen. "Reconceptualizing the 'Anthropos' in the Anthropocene: Integrating the Social Sciences and Humanities in Global Environmental Change Research." *Environmental Science & Policy* 28 (2013): 3–13.

Preston, Christopher J. "Ethics and Geoengineering: Reviewing the Moral Issues Raised by Solar Radiation Management and Carbon Dioxide Removal." *WIREs Climate Change* 4, no. 1 (2013): 23–37.

Sideris, Lisa H. "Science as Sacred Myth? Ecospirituality in the Anthropocene Age." In *Linking Ecology and Ethics for a Changing World*, edited by Richard Rozzi, S. T. A. Pickett, Clare Palmer, Juan J. Armesto, and J. Baird Callicott, 147–62. Dodrecht: Springer, 2013.

Steffen, Will, Jacques Grinevald, Paul Crutzen, and John McNeill. "The Anthropocene: Conceptual and Historical Perspectives." *Philosophical Transactions of the Royal Society of London A: Mathematical, Physical and Engineering Sciences* 369, no. 1938 (2011): 842–67.

Steffen, Will, Åsa Persson, Lisa Deutsch, Jan Zalasiewicz, Mark Williams, Katherine Richardson, Carole Crumley, Paul Crutzen, Carl Folke, Line Gordon, Mario Molina, Veerabhadran Ramanathan, Johan Rockström, Marten Scheffer, Hans Joachim Schellnhuber, and Uno Svedin. "The Anthropocene: From Global Change to Planetary Stewardship." *Ambio* 40, no. 7 (2011): 739–61.

Swimme, Brian, and Mary Evelyn Tucker. *Journey of the Universe*. New Haven: Yale University Press, 2011.

Tuana, Nancy, Ryan L. Sriver, Toby Svoboda, Roman Olson, Peter J. Irvine, Jacob Hagg-Misra, and Klaus Keller. "Towards Integrated Ethical and Scientific Analysis of Geoengineering: A Research Agenda." *Ethics, Policy & Environment* 15, no. 2 (2012): 136–57.

Chapter Nine

Ritual Responses to Climate Engineering[1]

Sarah E. Fredericks

As human failure to slow anthropogenic climate change has become apparent, some have begun to consider whether geoengineering—intentional, global manipulation of the climate—is a reasonable option. Philosophical and religious responses to this question tend to be rational, analytical investigations of geoengineering that list ethical issues it raises or the philosophical and theological reasons why geoengineering is problematic or possible given a particular vision of humanity. These are important analyses. Yet geoengineering is not merely a conceptual dilemma. It also requires ritual action, whether enacted in a traditional or emerging religion, to address the existential, emotional, relational, and physical issues it raises for individual humans, communities, and humanity as a whole.

To substantiate these claims, I proceed in five parts. First, I describe geoengineering and justify my focus on a particular type, solar radiation management. Second, I review the major ethical responses to geoengineering, arguing that although they emphasize different ethical content, they share deep assumptions about how ethics operates. Next, I argue that this common ethical response has a number of limits: it fails to recognize that people often fall short of their ideals, that existential questions raised by such failure demand responses, that emotions and relationships as well as rationality inform human actions, and that it needs a vision of human nature that allows hope of a better future. Fourth, I identify the characteristics of religious ritual that avoid the limits of existing ethical approaches to geoengineering. Finally, I suggest that ongoing geoengineering rituals within and between religions would be helpful since they deal with affective responses to geoengineering, facilitate reconciliation between a variety of groups, and enable action.

GEOENGINEERING

Geoengineering strategies include carbon capture and solar radiation management (SRM).[2] Carbon capture involves 1) capturing carbon dioxide in the atmosphere or made during fossil fuel use and 2) sequestering it, most likely underground. SRM increases the reflectivity of the atmosphere, buildings, or natural landscapes to decrease solar radiation absorbed by the atmosphere or earth and thus counter global warming. Carbon capture proposals are generally distributed systems (e.g., widespread tree planting) that will take decades or centuries to gain full effectiveness.[3] SRM techniques, especially the most popular plan of emitting sulfur into the atmosphere, will achieve results in a year or two and will have lower up-front costs. Thus, SRM is often seen as the most politically and technically feasible solution.[4] Yet it does not actually get to the heart of anthropogenic climate change (increased greenhouse gases) as it only addresses rising temperatures.[5] Additionally, SRM is a much more uncertain process than carbon capture and has potentially serious side effects. These include increased acid rain and decreased rainfall in "Asian and African summer monsoons," which in turn can decrease agricultural output. Furthermore, once started, it will likely be necessary to continue intentional sulfur emissions in order to avoid rebound effects worse than the pre-SRM temperatures.[6] Subsequently, SRM is the focus of most philosophical and religious responses to geoengineering, and it will be my focus as well, though I note that existing arguments against it should be sufficient to reject it.

ETHICAL RESPONSES TO THE DILEMMA OF GEOENGINEERING

Ethical analyses of geoengineering are still a rather new part of environmental ethics. While Dale Jamieson published one of the first articles on the subject in 1996, most of the literature dates from 2010 or later. It still tends to survey possible ethical issues in geoengineering rather than deeply examining one.[7] Despite the nascent state of the field, most climate engineering ethicists share common assumptions. They value slowing anthropogenic climate change and/or mitigating its effects. They assume that human degradation of natural systems is wrong and think that humans have a responsibility, duty, or obligation to right this wrong. Most oppose SRM, and sometimes any geoengineering, without more study and justification.[8]

First and foremost, critics worry that SRM's negative side effects will outweigh its benefits.[9] For instance, while it could rapidly cool the planet's temperature, its effect on rainfall patterns may be as bad as if not worse than that caused by climate change. Similarly, the risk for biota, ecosystems, and society seems too much for many analysts. Such judgments may rest on utilitarian calculations, concern for the intrinsic or instrumental worth of biota or people, duties to such entities, or a vision of the virtuous individual.

The principle of responsibility can yield similar conclusions. For some scholars, geoengineering may involve too much responsibility for humans to bear.[10] In part, this is due to its global scale, yet it is a distinct, and more difficult, moral issue than other global anthropogenic environmental changes including climate change, because geoengineering is intentional.[11] The inability to reverse some geoengineering methods including SRM without significant rebound effects only adds to the burden. It may rightly be overwhelming to assume the responsibility for the outcomes of an intentional, global, and irreversible action. Can one person, a small group, or even all people (if there was a way to come to meaningful consensus among so many) psychologically stand taking the blame if something goes wrong for intentionally manipulating the climate? Does such a group have the physical, political, economic, intellectual, and moral capacity to intervene successfully during or after geoengineering if something goes wrong? Without such abilities, it may be irresponsible to undertake geoengineering, especially the more risky versions of SRM.

Justice is another key concern in geoengineering ethics. Adam Corner and Nick Pidgeon recognize that minimizing the risks to various groups of people—especially the poor, politically powerless, and those most at risk geographically—is critical since such groups will be least able to deal with the negative side effects of geoengineering.[12] Relatedly, they (along with many others[13]) ask questions of procedural justice regarding geoengineering: Can the affected people give consent or participate in the decision-making process? Who gets to decide whether to start geoengineering and what kind to implement? How will the input of developing and developed nations, political leaders, business leaders, and citizens be included? Will future generations, who will likely experience the brunt of today's decisions about geoengineering, be considered? Who will provide the funding, the technical expertise, equipment, and the monitoring to implement and assess the effort? Who will be responsible (legally, financially, etc.) if geoengineering causes harms? Indeed, the Royal Society's report on geoengineering maintained that "the greatest challenge to the successful deployment of geoengineering may be the social, ethical, legal, and political issues associated with governance" and that international-scale geoengineering other than "removal of greenhouse gases from the atmosphere" should not take place without "appropriate governmental mechanisms."[14]

While the above ethical analyses emphasize different ideas (cost/benefit analysis, responsibility, justice), deeper methodological presuppositions unite them, allowing us to speak of a common ethical approach. The authors who have espoused this common ethical approach assume that they will be able to identify one or more ethical principles to judge whether it is permissible to pursue geoengineering or virtues to guide such decision-making. Little to no attention is paid to methods of implementing the ethic, how ethical recommendations get adopted and applied by communities, or even the fact that such a method may be needed. They presume that after such a ruling, people will simply implement it and the ethical issue will be solved.[15] Furthermore, potential problems implementing the ethical recommendations are rarely discussed, and if they are, authors presume that a reiteration of or recommitment to the original goal will solve the problem. This expectation should not be surprising, because assuming that people can and will follow through with ethical rulings is the mode of most modern Western ethics.

LIMITS OF THE COMMON ETHICAL RESPONSES TO GEOENGINEERING

The common ethical approach to geoengineering makes many strong points about the dangers of geoengineering. Yet, it is limited for a number of interrelated reasons: It is overconfident in the human capacity to live up to ideals and thus neglects to consider what to do when humans do fail. It neglects existential questions, emotions, and relationships as factors that influence decision-making and action along with rational analysis. It is often disconnected from visions of human nature such that it offers little to no hope for improved decision-making and action in the future. Let us examine each of these items in turn.

First, the common ethical approach neglects the fact that people often fail to live up to ethical ideals. It offers no way of dealing with failure except to analyze the new problem and issue a new judgment about what went wrong and what should be done next. Such a response ignores the root causes of failure. Competing alternatives, our limited knowledge, and the complexities of our social-ecological systems often mean there is no easy answer. We often have limited capacities to enact our plans whether due to a weak will, fear of the unknown, or the tendency to overlook issues that are insidious, slow-moving, and on a large spatial scale (i.e., climate change).[16] Significant emotional, social, political, and infrastructural conditions often hinder changes on the individual and social levels. Ignoring these contributing factors to failure means that our plans about how to achieve our ethical ideals are insufficient.

Certainly, some climate change ethicists, most notably Stephen Gardiner and Dale Jamieson, recognize some of these limitations in ways that presumably apply to an analysis of geoengineering as well.[17] Indeed, Stephen Gardiner helpfully names the tendency to involve ethical systems and then assume them to be straightforwardly and sufficiently applied the "invoke and apply" model.[18] However, the alternatives offered typically follow the same old pattern or are insufficient. For instance, Gardiner recognizes that many factors including power, financial pressures, the allure of the status quo, assumptions of prior entitlements, and emotions including resentment exacerbate the moral challenges of climate change and geoengineering. I applaud his efforts to unpack the realities of the complex moral situation in which we find ourselves instead of just emphasizing ideals. Yet I disagree that when faced with these potential contributors to moral corruption, one primarily needs to double down on rationality and clear arguments.[19] Rather, I wonder whether human limitations mean that rationality is an imperfect tool and that assuming it is perfect can cause additional problems. A fuller ethical system would acknowledge and work within human limits and their psychological, existential, and relational effects.

Dale Jamieson also emphasizes rationality, as the title of his recent climate ethics book, *Reason in a Dark Time*, indicates. Admittedly, Jamieson acknowledges that climate change is beyond our affective capabilities,[20] that people may be affected by despair,[21] and that some will feel "grief, guilt, and nostalgia" when confronted with the fact that climate change was caused by us.[22] Yet the majority of his text examines how traditional ethical systems or his own proposals will deal with cognitive limitations rather than emotion or will. For instance, he recognizes the limits of our ethical theories of collective and individual responsibility. He also notes that as humans we evolved to respond to mid-sized objects moving quickly where causes are linked closely in time and space to effects rather than the slow, diffuse, processes of climate change.[23]

Jamieson's approach differs from Gardiner's because he calls for a turn to virtue instead of a principled or rule-based ethic. This move may circumvent the common ethical response's tendency to point to the rules when failure occurs since virtue ethics allows for a process of development in which a person may move toward or away from becoming virtuous. Yet Jamieson does not sufficiently explore whether and how the process of cultivating virtues is affected by climate change in general or geoengineering in particular. Thus, he does not yet have a means of implementing his ethic that deals with the significant limitations of humanity in trying to live up to their ethical vision given the conceptual, practical, emotional, and existential challenges of geoengineering.

Certainly, Gardiner and Jamieson identify some of the impediments to ethical decision-making about climate change, including geoengineering, and both helpfully push the conversation forward. Yet the capacity for ethical action does not come entirely from rational arguments that can stand up to our cognitive limitations or the potential moral corruption of climate change.

Humans are not merely rational creatures. We also have emotions that can significantly shape our perceptions, assessments, and desires. Ignoring emotions to merely reiterate the rules, or even, as Gardiner and Jamieson do, to analyze the complexities of the situation and then return to rules or virtues, is insufficient. These approaches do not adequately deal with how to do better the next time because they do not sufficiently explore the emotional, existential, and social consequences of failure.

The conditions that lead people to consider geoengineering reveal the irony of trying to respond to it with such a limited ethical toolkit. Geoengineering is only being considered because humans recognize that their past actions have had significant unintended consequences and yet have continued to pump greenhouse gases into the atmosphere well past the point at which many knew this was problematic. We (or rather some people) are considering geoengineering because we have tremendous physical power as a species stemming from our intelligence, yet this intelligence is not complete. We did not understand what we were doing with respect to the extraordinarily complex earth system when we started getting ourselves into this mess. Then, even when we began to understand, we lacked the trust in science, the political capacities, and individual and social will to buck the trends of capitalism and short-term luxury to take quick action. By contributing to anthropogenic climate change, humans, as a species, have changed the global system in ways that will have decades if not centuries or millennia of effects on individual organisms, communities, species, and ecosystems.[24] We are threatening our civilization as we know it.

As Willis Jenkins notes in the previous chapter of this volume, the scale and severity of the threat that we have brought upon ourselves raises existential questions that the ethics described above cannot deal with: How could we do this? What kind of beings are we to bring about such destruction? How can we expect to change the climate enough to counteract our emissions without also creating severe negative consequences? These questions imply an assessment of human nature not enabled by the mere judgment of right or wrong or analysis of the social causes of ethical dilemmas so common in contemporary ethics. Fortunately, philosophers, ethicists, and theologians reflecting on geoengineering have raised these questions too, though as we shall see, their responses are often too conceptual and/or abstract to address them fully.

Indeed, one of the most common objections to geoengineering sees it as assuming too much power or going against the natural order.[25] Often, scholars summarize these arguments as hubris, overstepping human bounds or capacities. For some, this argument is rooted in theological beliefs about the relative power of God and humanity.[26] For most other authors evaluating geoengineering, these objections are not explicitly rooted in religion (few religious scholars have yet written on geoengineering), though they may use religious imagery. Benjamin Hale writes, "Manipulating the earth's climate essentially places human geoengineering in the role of gods, and as such, there is a hazard that these engineers will come to believe themselves to maintain control of the climate. The moral hazard is engendered in instigating the vice of hubris, of encouraging us to adopt further approaches in a way unbecoming of a good person."[27] Scholars may also maintain that it is "wrong on principle to meddle intentionally with something as fundamental as solar radiation."[28]

These concerns point to the existential questions of humanity that are insufficiently addressed by the standard ethical response. Existing analyses of hubris, for instance, focus on a narrow aspect of what it means to be human—as a thinking thing. Certainly describing the rational, knowledge-lacking elements of hubris and whether it may lead to "self-aggrandizement" are important issues in their own right.[29] If one stops there, however, it is too easy to fall into the old pattern of assuming that describing the problem will necessarily lead to the right analyses and ethical principles that in turn will be perfectly implemented. The solution is not just to have different, better knowledge, or even to believe differently (i.e., not to have a big head about one's own power), but to have practices and community structures that reinforce such different beliefs and, just as importantly, encourage and enable action in spite of human limitations.

The common ethical approach is also problematic because it ignores the way that failure can trigger emotions of negative self-assessment that can then exacerbate the original problem. Moral psychologists recognize that when people feel shame (a negative self-judgment about one's whole self because one has done an action one judges to be wrong, as opposed to guilt, a judgment about the action itself) they often retreat, hide, deny the problem, or get angry rather than apologizing or taking action to mitigate the problem or prevent it from recurring.[30] Studies by Mark A. Ferguson and Nyla R. Branscombe,[31] surveys,[32] environmental memoirs and self-help books,[33] and my own work on online confessions of environmental wrongdoing[34] indicate that increasing numbers of people are feeling shame about contributing to environmental destruction and climate change.[35] The number of people feeling environmental shame is likely to increase as more people 1) adopt environmental values and 2) recognize the extent of the threat of climate change and their contributions to it. Geoengineering can exacerbate such feelings in at least two ways. Existential questions raised by geoengineering are often answered negatively by opponents of geoengineering who often see humanity as a blight to or problem in the world. These judgments parallel the conditions of shame. Secondly, if SRM is implemented and has significant negative consequences, as is likely, decision-makers and those who identify with them are likely to have significant negative self-assessments due to the monumental scale of the problem, the difficulty in reversing it without rebound effects, and the intentionality of their actions. Of course, shame is problematic as it sparks denial and retreat rather than necessary action.

Additionally, the common ethical approach usually does not acknowledge let alone deal with the interpersonal and societal consequences of failing to live up to ethical ideals. If SRM is implemented, it is likely to have the serious side effects discussed earlier (e.g., food shortages due to weather changes) as well as indirect but also potentially serious effects on individuals and society including as yet unknown psychic effects of sky whitening and the disruption of socially-significant agricultural practices. People least able to cope with such physical, psychic, economic, and societal changes will likely disproportionately experience them. Furthermore, people adversely affected by SRM will probably know that they were harmed by intentional decisions by people who did not value them highly enough to prevent such harms. These experiences are problems in and of themselves because they fly in the face of human dignity and because such stresses can further affect physical and psychological health and community vitality. They may impede future policy-making and action as affected people distrust decision-makers and get increasingly angry about geoengineering injustice.

Restorative justice scholars and activists discuss similar issues[36] but rarely with respect to climate change,[37] and none that I have seen with respect to geoengineering. This is not surprising, for restorative justice analyses and initiatives typically occur after an injustice and SRM has not yet been deployed on a large scale. If geoengineering proposals are to be evaluated according to their impact on humans, as they should be, then psychological and community well-being also need to be considered before and during, not just after its implementation.

The common ethical systems are also problematic because they offer no comfort, no support, and no hope that someone may do better next time. Reiterating ethical reasoning may not help someone struggling to be ethical. Pairing reasons with a vision of humanity that enables change, methods for making such changes, and social support through the changes are often much more successful ways to bring about change, as gym buddies and study partners recognize. Thus, ethical approaches to geoengineering need to include more than ideals of what should happen; they also need to include methods of implementation. We need an approach that can deal with the rational as well as emotional and existential issues of individuals and communities as they make decisions and truly *act* together in ways that acknowledge human limits and provide hope and possibilities for the future. Religious rituals share all of these features.[38]

RELIGIOUS RITUAL

To understand how religious ritual is a helpful response to geoengineering, I first need to clarify the terms religion and ritual. Religion includes phenomena popularly conceived of as religious—that is, of Judaism, Hinduism, and so forth—as well as phenomena of new religious movements, civil religions, nature religions, and other movements. They generally share characteristics including doctrines, rituals, narratives, emotions, ethics, social structures, and material objects[39] though, contrary to common Western stereotypes, they may not include belief in a personal deity or use of a sacred text.

Turning to ritual, I follow Catherine Bell's and Jonathan Z. Smith's work. Bell notes that in the modern West, rituals are often conceived of as incredibly formal, traditional, governed by rules, and static. They are thought to deal in sacred symbols and connect practitioners to the divine or transcendent.[40] Yet, Bell rightly argues that rituals actually include a much broader array of actions. Ritual is at heart "a way of acting to differentiate some actions from others" rather than any particular set of special practices.[41] For Smith, ritual is "a mode of paying attention . . . a process for marking interest" not of blind action or habit, but of attentiveness.[42]

While "paying attention" may suggest cognitive processing, rituals involve much more: movement, the senses, and emotion. Rituals do not just play out preconceived notions or enable thought. They also spark new experiences of the body, emotions, and relationships to others. Rituals allow participants to pay attention to or connect to experiences, things, place, co-ritualists, and, possibly, the transcendent. Indeed, ritualization makes meaning.

Regardless of the overt focus of rituals, they create and maintain the identity of participants as individuals and in relation to other ritual participants. Identity often develops in contrast as participants differentiate themselves from their preritual selves or the people who do not participate in the ritual. Communities support ritual participants in their belief and practice, such that an individual's identity is bound up with that of their fellow participants. Insofar as the rituals deal with fundamental issues of humanity, identity forged in ritual can be species identity.

Ritual differentiation also occurs between the ordinary and the ideal. Whether dancing, singing, or performing other movements, rituals identify what is ideal and enact it in some way.[43] Think, for example, of a holiday on which people try to put aside their differences to have a family meal. The ritual lets participants articulate the ideal and recognize their typical distance from it. Afterward they have a goal to strive for in their everyday lives. Thus, people acknowledge their limits, whether a particular problematic action, limited relationship with others, or even the existential limitations of self or species, through ritual.

A specific type of ritual, "rites of affliction," which "attempt to rectify a state of affairs that has been disturbed or disordered,"[44] is particularly relevant to geoengineering. Many different religious traditions have such "rites of affliction" [45] that acknowledge human limits and articulate ways to deal with them and restore hope. For instance, Baren J. de Klerk argues that traditional Christian liturgies recognize the need for lament, grief, and confession as we humans continually fail to live up to ideals. Yet he notes that the service also praises and hopes as it continually renews the possibility of something better.[46] This tension ideally deals with the negative emotions associated with failure, guilt, loss, and pain without wallowing in it. Here—as well as on the Day of Atonement in Judaism or during processes of relinquishing desires or perceived wrongs to oneself in Buddhism—participants acknowledge particular and existential wrongs, maintain the worth of participants, and chart a path forward in which the harm should decrease. The existential explanation for pain is not just thought about, it is understood through embodied experience. These existential understandings infuse the sense of self, community, and possibly species identity that emerges through the ritual.[47]

RITUALS FOR GEOENGINEERING

Geoengineering requires multiple intertwined types of rectification because it involves multiple intertwined types of disorder (physical, relational, emotional, existential). The need to respond physically to environmental crises demands movement. Physical acts, say planting a tree or launching planes to release sulfur, can materially respond to climate change and symbolize larger initiatives. Movement in rituals may also include entirely symbolic gestures of repairing climate systems—for example, a skit or song about alternative energy—which help transform participants' sense of self, community, and the possible and subsequently may lead to changed action. Relational rituals, particularly of confession, apology, and atonement, can begin to rectify the emotional, existential, and relational disturbances of geoengineering. Confessing wrongdoing or failing to live up to one's ideals can call attention to the distance between ideal and actual self or community and an opportunity to reinforce or develop a vision of humanity in which there is a difference between one's action and identity such that positive change is possible. They can provide a temporal break after which one can rededicate oneself anew to the ideal. All of these moves can help reconcile an individual or community to him/her/themselves by dealing with their emotional and existential response to their limitations. Such initiatives can also repair or generate relationships between perpetrators and victims, whether individual or collective, as each acknowledges and tries to understand the problems of past, present, and potentially future and works together to decrease future harm. While such restorative rituals are usually conceived of occurring between humans currently alive, some may explore symbolic or actual ways to reconcile with nonhuman entities in ecosystems, future generations of humans, or the divine. Of course, reconciliation with ultimate reality/ies may also be important for particular religious groups; these rituals may include modifications of traditional rituals devoted to this purpose.[48]

Despite the fact that rituals can address so many types of disorder related to geoengineering, few rituals explicitly target geoengineering. The work of the IASS, described in chapter 2, is a rare example to which I return in a moment. First, I look to the more prevalent rituals which deal with the fallout of anthropogenic environmental degradation and could serve as exemplars for geoengineering rituals since geoengineering-specific rituals are still so rare. For instance, many Christian groups have come to recognize harm to the environment as a sin and worked it into their confessions. William Jordan, a restoration ecologist, and Gretel Van Wieren, a Christian environmental ethicist, argue that restoring ecosystems is a ritual process. Restoring a prairie can restore the physical conditions of the ecosystem, repair the relationship between actors and environment, and deal with shame about the fact that to be a human is to be involved in destruction, at minimum because we need to eat.[49] Paul Santmire advocates more deeply integrating nature into Christian liturgies and extraliturgical rituals such as tree planting ceremonies and the combination of stations of the cross and cosmos.[50] Scientist Chan Eng Heng developed new rituals using religious themes to protect turtles in Malaysia.[51]

Ritualization also happens outside of traditional religious communities. Johanna Macy and colleagues have developed a series of exercises drawing on Buddhism, systems theory, and deep ecology to help participants deal with grief and fear prompted by environmental destruction.[52] Bronislaw Szerszynski identified the ritualistic characteristics of environmental protest movements in England: protestors aim to differentiate between the world as it is and how they wish it to be, create new meanings of what is ideal, develop a connection between the specific and universal, and forge identity among themselves.[53] Climate action groups such as 350.org may embody similar characteristics. I documented an emerging online ritual in which environmentalists confess their guilt, deal with it and their shame, and share ideas of how to do better.[54] Restorative justice initiatives reconcile polluters and communities whose environments and quality of life degraded by the pollution.[55]

These ritual initiatives can serve as a model for responses to geoengineering because they address issues critical to geoengineering including emotions, existential questions, community relationships, ethical ideals, and the need to take concrete physical and policy action to mitigate or reverse environmental damage and/or prevent future harms. Yet none of these rituals is sufficient to deal with geoengineering. They have a limited scale of reconciliation and action compared to the global propositions of SRM. Most deal with participants in a small community at one time, an insufficient move when a small group of people in one place may initiate SRM in other places and impact people all over the world now and into the future.

Though existing rituals will not immediately address the needs of geoengineering, I (or any scholar) cannot simply construct an alternative ritual and expect it to catch on. Rituals must have meaning for and arise out of the communities that practice them. Indeed, people generally consider rituals without connection to tradition "anomalous, inauthentic, or unsatisfying."[56] Yet rituals are not necessarily static. Religious communities often extend old rituals and theological categories to cover new situations; some confessional rituals now include environmental sins. Other religious rituals related to the environment may resonate with the tradition—for example, creation care—but involve a newer ritual practice—for example, planting trees. Similarly, "secular" rituals—whether of protest movements, online confessions, or restorative justice initiatives—build upon themes and practices familiar to their participants. Thus, new rituals can be developed, as in the work of the IASS and in the workshops and books of Joanna Macy and her coauthors. Macy's work aims to deal with the grief and lament of environmental destruction and spark hope for a future in which participants enact a better world.[57] Her practices and other emerging rituals catch on because they fill needs of the people, resonate with practices that are familiar, and are revised over time to better meet and reflect these needs.

To balance the need for the familiar with the challenges of scale, I propose two levels of geoengineering ritualization and then discuss their common features.[58] The first develops rites of affliction regarding geoengineering among a group (a neighborhood religious group, regional or national religion, or even global religious community) with shared experiences of reconciliation, atonement, forgiveness, detachment, and so forth. At this level, participants can extend what they already know or do through their traditional processes to face new challenges through ritual.

The second level involves intentional partnerships between two or more communities with significantly different contributions to and experiences of climate change and geoengineering and most likely, different political, economic, and religious backgrounds.[59] Intimate, prolonged exchanges (probably online to avoid the monetary and environmental expense of physical travel) with each other allow the depth of exchange needed to apologize as necessary and forgive as appropriate. Interactions with people outside of one's tradition and place can symbolize the need to reconcile with all humans before, during, and after geoengineering. Paying attention to the broader impacts of such partnerships and sharing their results with a wider audience will help move past a merely symbolic global interaction. Participants can also work together to stop problematic actions in the future and physically, socially, and financially mediate geoengineering's negative effects. Such collaboration can also aid existential insights by exposing participants to a perspective lacking in their own tradition or helping them come to appreciate the benefits of their tradition through comparison.[60] Of course, cross-religious ritualization brings its own challenges because it can be difficult to respond to new challenges (e.g., geoengineering) while resonating with all of the traditions involved. Existing collaborative rituals of Earth Day, restorative justice, and ecumenical and interfaith activities for social justice or memorial services after a tragedy suggest that such interaction is possible, though challenging.

While these levels are different, they can share several features. Rituals of reconciliation at either level can heal or establish many kinds of relationships. They can reconcile you to yourself (as you cope with your contributions to the need for geoengineering and the process of doing it) or to others in your community (at the first level) and outside of it (at the second level). Rituals should acknowledge those who may have been harmed by your contributions to climate change or geoengineering or who may have harmed you with their contributions. Intergenerational reconciliation poses unique challenges. Whether it involves the dead or unborn, careful intentional attention to one's emotions can help lead to either the forgiveness of ancestors or public apology and a commitment (with follow through) to changing conditions for the future. Individuals and communities can also potentially reconcile with nonhumans as people recognize their unique nature, influence on humans, and own needs.[61] Finally, individuals and the community as a whole may be reconciled to the ultimate or ultimates as conceived in their religion through offerings, grace, or the relinquishing of desires.

Geoengineering rituals should also serve as a space to acknowledge people's affective response to geoengineering, whether anger, grief, guilt, shame, or despair that others (and/or themselves) put them in the position to consider it and/or deal with its side effects. Rituals can allow people to feel the depth of the problem without wallowing. People need the hope of an alternative individual and community identity that recognizes human power and limits and can arise in rituals.

Action, whether symbolic or real, whether through political protest, education, economic pressures, or physical changes, is also needed for physical reasons and because it is critical to the most effective apologies and subsequently, to correct injustices. Yet perfection should not be the only measure of the ritual's success or else we have failed to escape from the ethical approach discussed earlier.

Rituals of affliction need to be ongoing or repetitive in order to cope with past and probable future discrepancies between ideal and actual actions. Repeating rituals rather than having one-off events also responds to the ongoing nature of geoengineering: it is born of a desire to address past wrongs and problems, and its effects will reverberate across the globe, through the atmosphere, biota, and human psyches. An extended process of ritualization also enables people to build and maintain relationships so that they can empathize with, honor, apologize, or forgive the other and self as necessary and try to change their actions. These things do not happen overnight. Finally, ritualization over time gives participants the opportunity to revise the ritual to be more meaningful and psychologically, politically, and physically effective.

Speaking of time, different phases of geoengineering require different rituals. While studying geoengineering and considering whether and how to implement it, rituals can help people deal with the magnitude and gravity of what brought them to consider geoengineering, their decision to act in spite of its probable side effects, the difficulty or impossibility of reversing it, and its implications about human nature. The activities of IASS Potsdam described in chapter 2 are one example of rituals that occur before geoengineering is implemented. Their discussions focused on an exchange of information so that participants from the religious groups could learn about technical possibilities and limits of geoengineering, and technical leaders and policy experts could learn about the moral reactions of the religious groups and individuals. Throughout the events, IASS organizers took care to make people feel welcome and to foster communication and trust between participants through the physical setup of the meeting (e.g., meeting at a round table) as well as its schedule (e.g., time was set aside to share personal responses to the idea of geoengineering). Indeed personal reactions and feelings were so valued in and of themselves and a part of building trust in the community that organizers made sure everyone had a chance to share even if it meant pushing back later items on the prearranged agenda. Through this setup, presentations about geoengineering by IASS members, and lots of time for participants to talk while IASS members listened, the meetings facilitated the exchange of information and a deeper awareness of geoengineering. These events are an early piece of the pre-geoengineering ritual process as IASS members wisely suggest needs to happen even before decisions or recommendations are made. Though this process is only a narrow part of what may be needed to fully evaluate geoengineering, in it we do see a deep recognition of the importance of community, relationships, affect, and ethics even among scientific and policy experts, a signal of the importance of these features in a full response to geoengineering.

If SRM is implemented, other rituals will also be needed continually during it, since SRM, particularly emitting sulfur, is not a one-time event. Rituals during implementation may, for example, ritualize the process of sending planes to spread sulfur. They may also involve reflection about doing geoengineering—to foster hope and deal with fear. Hope is especially necessary in the period between deploying a strategy and seeing its effects. Participants also need ways to psychically and/or procedurally deal with the fear that it will not work. They may cognitively and existentially recognize human characteristics (political, financial, will, emotion, etc.) that may help or hinder the process. They should ponder, assess, and respond to hubris fostered by the SRM experience and its implications for other projects. They should recognize injustices from SRM and to work to lessen their effects. They should determine responsibility for the decision to undertake SRM and who takes the blame or credit for its consequences.[62] Implementation rituals should also build identity about maintaining SRM processes to convince others to fund and maintain the project, especially in new generations to avoid rebound effects. Additionally or alternatively, implementation rituals should enable reassessments of the project, and if it is deemed no longer acceptable, find ways to combat rebound effects if SRM ever needs to be or is stopped. These rituals will help people deal with its actual rather than predicted effects.

If SRM is stopped, we will also need rituals to assess how it worked physically, relationally, emotionally, ethically and deal with its outcomes in all of these dimensions. Furthermore, reflection on how humanity got to the place of success, failure, or mixed results and what can be learned about the human condition, if anything, through these actions will be needed to make sense of decisions to pursue geoengineering if in fact such decisions are made.

Admittedly, these features of rituals for geoengineering are vague and probably form an incomplete list.[63] Specificity will come as people develop rituals, a project beyond the scope of this chapter. Here, I aimed to show that rituals respond to challenges of geoengineering not met by common ethical approaches. Geoengineering's intentionality and global scale in the face of anthropogenic climate change necessarily call into question human possibilities and challenge our emotions, relationships, and physical actions. Rituals can respond not only by articulating ideals, as ethics commonly does, but also by addressing the limits and possibilities of human existence through action with others. Geoengineering demands this multifaceted response.

NOTES

1. Thanks to the other contributors in this volume, especially Forrest Clingerman and Kevin O'Brien, for their helpful comments. Thanks also to the participants in the Theology and Ethics Workshop at the University of Chicago, especially Willa C. Swenson-Lengel, for their questions.

2. For more information about geoengineering, see the introduction to this volume and The Royal Society, *Geoengineering the Climate* (London: The Royal Society, 2009); Christopher J. Preston, "Solar Radiation Management and Vulnerable Populations," in *Engineering the Cli-*

mate: The Ethics of Solar Radiation Management, ed. Christopher J. Preston (Lanham, MD: Lexington Books, 2013).

3. The Royal Society, *Geoengineering the Climate*, 8.

4. Stephen M. Gardiner, "Some Early Ethics of Geoengineering the Climate: A Commentary on the Values of the Royal Society Report," *Enviromental Values* 20 (2011): 168; Clive Hamilton, "The Ethical Foundations of Climate Engineering," http://clivehamilton.com/the-ethical-foundations-of-climate-engineering/; Dale Jamieson, "Ethics and Intentional Climate Change," *Climatic Change* 33, no. 323–36 (1996): 325; Preston, "The Extraordinary Ethics of Solar Radiation Management," 2–3.

5. Marion Hourdequin, "Geoengineering, Solidarity, and Moral Risk," in *Engineering the Climate: The Ethics of Solar Radiation Management*, ed. Christopher J. Preston (Lanham, MD: Lexington Books, 2013), 23.

6. The Royal Society, *Geoengineering the Climate*, 31; A. Robock, "20 Reasons Why Geoengineering May Be a Bad Idea," *Bulletin of Atomic Scientists* 64, no. 2 (2008): 15–17.

7. Jamieson, "Ethics and Intentional Climate Change"; Wylie Carr, Ashley Mercer, and Clare Palmer, "Public Concerns about the Ethics of Solar Radiation Management," in *Engineering the Climate: The Ethics of Solar Radiation Management*, ed. Christopher J. Preston (Lanham, MD: Lexington Books, 2013); Hamilton, "The Ethical Foundations of Climate Engineering"; Ben Hale, "The World That Would Have Been: Moral Hazard Arguments against Geoengineering," in *Engineering the Climate: The Ethics of Solar Radiation Management*, ed. Christopher J. Preston (Lanham, MD: Lexington Books, 2013).

8. See the introduction of this volume for a discussion of the differences between adaptation, mitigation, and geoengineering.

9. Carr, Mercer, and Palmer, "Public Concerns about the Ethics of Solar Radiation Management," 173–74; The Royal Society, *Geoengineering the Climate*; Robock, "20 Reasons Why Geoengineering May Be a Bad Idea."

10. Preston, "The Extraordinary Ethics of Solar Radiation Management," 5; Hamilton, "The Ethical Foundations of Climate Engineering," 18.

11. Preston, "The Extraordinary Ethics of Solar Radiation Management," 10, 17–18; Adam Corner and Nick Pidgeon, "Geoengineering the Climate: The Social and Ethical Implications," *Environment* 51, no. 1 (2010); Hamilton, "The Ethical Foundations of Climate Engineering".

12. Corner and Pidgeon, "Geoengineering the Climate: The Social and Ethical Implications," 29–30.

13. Carr, Mercer, and Palmer, "Public Concerns about the Ethics of Solar Radiation Management," 179–80; Jamieson, "Ethics and Intentional Climate Change," 329.

14. The Royal Society, *Geoengineering the Climate*, xi. Yet, as Gardiner points out, the report actually spent little time on such issues (Gardiner, "Some Early Ethics of Geoengineering the Climate: A Commentary on the Values of the Royal Society Report"). The report emphasized the technical issues that are its expertise. Ideally, experts on governance would produce and enact a companion; though once again policy lags technical exploration.

15. Stephen M. Gardiner, *A Perfect Moral Storm: The Ethical Tragedy of Climate Change* (New York: Oxford University Press, 2011), 308.

16. Dale Jamieson, *Reason in a Dark Time: Why the Struggle against Climate Change Failed—and What It Means for Our Future* (New York: Oxford University Press, 2014).

17. Gardiner, *A Perfect Moral Storm: The Ethical Tragedy of Climate Change*; Jamieson, *Reason in a Dark Time: Why the Struggle against Climate Change Failed—and What It Means for Our Future*. Admittedly, their work emphasizes climate change over geoengineering, but their ideas apply to geoengineering as well.

18. Gardiner, *A Perfect Moral Storm,* 308.

19. Gardiner, *A Perfect Moral Storm,* 337, 399–407.

20. Jamieson, *Reason in a Dark Time: Why the Struggle against Climate Change Failed—and What It Means for Our Future*, 8.

21. Ibid., 178.

22. Ibid., 238.

23. Ibid., 4.

24. Yes, I know that some people have been much more responsible for climate change than others and some are certainly disproportionally affected, but all humans have contributed since it is impossible to divide carbon dioxide molecules emitted from cooking open fires and from luxury vehicles; the combined effects of our actions as individuals and nations cause the problem.

25. Hourdequin, "Geoengineering, Solidarity, and Moral Risk," 23; Preston, "The Extraordinary Ethics of Solar Radiation Management," 5; Hale, "The World That Would Have Been: Moral Hazard Arguments against Geoengineering," 127; Carr, Mercer, and Palmer, "Public Concerns about the Ethics of Solar Radiation Management," ibid., 176; Forrest Clingerman, "Between Babel and Pelagius," in *Engineering the Climate: The Ethics of Solar Radiation Management*, ed. Christopher J. Preston (Lanham, MD: Lexington Books, 2013), 209; Hamilton, "The Ethical Foundations of Climate Engineering" 10–11; Ben Minteer, "Geoengineering and Ecological Ethics in the Anthropocene," *BioScience* 62, no. 10 (2012); Northcott, *A Political Theology of Climate Change*, (Grand Rapids, MI: Wm. B. Eerdmans Publishing Co., 2013), 270.

26. Michael S. Northcott, *A Moral Climate: The Ethics of Global Warming* (Maryknoll, NY: Orbis Books, 2007); Clingerman, "Between Babel and Pelagius," 209; Forrest Clingerman and Kevin J. O'Brien, "Playing God: Why Religion Belongs in the Climate Engineering Debate," *Bulletin of Atomic Scientists* 70, no. 3 (2014).

27. Hale, "The World That Would Have Been," 127.

28. Preston, "The Extraordinary Ethics of Solar Radiation Management," 5.

29. Hale, "The World That Would Have Been: Moral Hazard Arguments against Geoengineering," 127; Carr, Mercer, and Palmer, "Public Concerns about the Ethics of Solar Radiation Management," 176.

30. June Price Tangney and Ronda L. Dearing, *Shame and Guilt* (New York: The Guilford Press, 2002), ix, 24–25.

31. Mark A. Ferguson and Nyla R. Branscombe, "Collective Guilt Mediates the Effect of Beliefs about Global Warming on Willingness to Engage in Mitigation Behavior," *Journal of Environmental Psychology* 30, no. 2 (2010).

32. National Geographic and GlobeScan, "Greendex 2012: Consumer Choice and the Environment—a Worldwide Tracking Survey," (Washington, DC: National Geographic and GlobeScan, 2012), 77; National Geographic, "Greendex: Consumer Choice and the Environment—a Worldwide Tracking Survey," National Geographic, http://environment.nationalgeographic.com/environment/greendex/?rptregcta=reg_free_np&rptregcampaign=20131016_rw_membership_r1p_us_se_w#; Lexmark International, Inc., "Consumers Concerned, yet Clueless, on 'Green' Technology," Lexmark International, Inc., http://newsroom.lexmark.com/index.php?s=13630&item=23708; Shelton Group, "Americans' Biggest 'Green' Guilt: Wasting Food, Survey Finds" (Knoxville, TN, 2012); Cone Communications, "CSR Press Release: All That Glitters Isn't Gold: This Year, It's Green," (2007), http://www.csrwire.com/press_releases/19356-All-That-Glitters-Isn-t-Gold-This-Year-It-s-Green; Maryellen Mooney and Gina Stikes, "Ivillage and Today.Com 'Eco-Confessions' Survey," Most Parents Admit They Could Be Greener," http://www.nbcnews.com/id/47064299/ns/about-press_releases/#.Ut2KwxDnaM9; Andrew Eisner to Retrevo Blog, 2013, http://www.retrevo.com/content/blog/2013/04/earth-day-2013-special-gadgetology-report; Kyle Freman, "Survey: Does Eco Guilt Stop You from Shopping at Ikea?," Apartment Therapy, http://www.apartmenttherapy.com/survey-does-eco-64103; Wendy Koch, "Going Green, Women Show More Eco-Guilt in New Year's Resolutions," Web, *USA Today: Green House* (2009), http://content.usatoday.com/communities/greenhouse/post/2009/12/going-green-women-show-more-eco-guilt-in-new-years-resolutions/1#.Ut1ifRDnaM-; Graceann Bennett and Freya Williams, "Mainstream Green: Moving Sustainability from Niche to Normal," in *The Red Papers* (Chicago: Ogilvy & Mather, 2011).

33. See, for instance, Fritzie von Jessen, *I Killed a Penguin: An Ecological Memoir* (Baltimore, MD: Publish America, 2009); Jen Pleasants, *Bag Green Guilt* (United States: Show the Love Media LLC, 2009); Prashant Vaze, *The Economical Environmentalist: My Attempt to Live a Low-Carbon Life and What It Cost* (London: Earthscan, 2009).

34. Sarah E. Fredericks, "Online Confessions of Eco-Guilt," *Journal for the Study of Religion, Nature and Culture* 8, no. 1 (2014).

35. Many of the surveys actually use the term guilt rather than shame, but it is widely documented that the general public often has a hard time differentiating between the emotions and that academic definitions differ significantly as well (Tangney and Dearing, *Shame and Guilt*, 10–12). Thus, if a survey just asks if people feel guilty about their environmental activities, it is not entirely clear if they feel guilt or shame according to Tangney and Dearing's definitions. At least some of the time such respondents feel Tangney and Dearing's shame and not guilt as indicated by their withdrawal, anger, or judgment of their whole self as problematic.

36. Robert Melchior Figueroa and Gordon Waitt, "Climb: Restorative Justice, Environmental Heritage, and the Moral Terrains of Uluru-Kata Tjuta National Park," *Environmental Philosophy* 7, no. 2 (2010); Sarah M. Conrad, "A Restorative Environmental Justice for Prison E-Waste Recycling," *Peace Review* 23, no. 3 (2011).

37. A few of the exceptions include Maxine Burkett, "Climate Reparations," *Melbourne Journal of International Law* 10 (2009); Stanley Kwenda, "World: North Should Pay South Reparations for Climate Change," *Inter Press Service*, May 14, 2010; Sasha Chavkin, "Cash for Thunder: Bolivia Demands 'Climate Reparations,'" *Mother Jones* (2009); Simon Caney, "Environmental Degradation, Reparations, and the Moral Significance of History," *Journal of Social Philosophy* 37, no. 3 (2006).

38. Surprisingly, Clingerman and O'Brien, who do examine what religion can bring to geoengineering, do not talk about ritual in "Playing God: Why Religion Belongs in the Climate Engineering Debate."

39. Ninian Smart, *Dimensions of the Sacred: An Anatomy of the World's Beliefs* (Berkeley: University of California Press, 1996).

40. Catherine Bell, *Ritual: Perspectives and Dimensions* (New York: Oxford University Press, 1997), 139.

41. Jonathan Z. Smith, *To Take Place: Toward Theory in Ritual*, ed. Jacob Neusner, William Scott Green, and Calvin Goldscheider, Chicago Studies in the History of Judaism (Chicago: The University of Chicago Press, 1987), 103.

42. Ibid.

43. Ibid., 109.

44. Bell, *Ritual: Perspectives and Dimensions*, 115.

45. Bell, *Ritual: Perspectives and Dimensions*, 115.

46. Barend J. de Klerk, "Enhancing Ecological Consciousness through Liturgical Acts of Doxology and Lament," *Verbum et Ecclesia* 35, no. 2 (2014).

47. Notably, not all contemporary Christians maintain this liturgical tension. Many do away with lament or even confession of sins in the interest of time or to avoid negativity.

48. Since rituals can involve movement, relationships, experiments, and emotions, they can also counteract the tendency of many ethicists to overemphasize analysis as we attempt to solve problems. See chapter 10 of this volume for more on the importance of repentance while considering geoengineering.

49. William R. Jordan III, *The Sunflower Forest: Ecological Restoration and the New Communion with Nature* (Berkeley, CA: University of California Press, 2003); Gretel van Wieren, *Restored to Earth: Christianity, Environmental Ethics, and Ecological Restoration* (Georgetown: Georgetown University Press, 2013).

50. H. Paul Santmire, *Ritualizing Nature: Renewing Christian Liturgy in a Time of Crisis* (Minneapolis: Fortress Press, 2008), 65, 67, 101, 78.

51. Michael Northcott, "Buddhist Rituals, Mosque Sermons, and Marine Turtles: Religion, Ecology, and the Conservation of a Dinosaur in West Malaysia," *Journal for the Study of Religion, Nature & Culture* 6, no. 2 (2012).

52. Joanna Macy and Chris Johnstone, *Active Hope: How to Face the Mess We're in without Going Crazy* (Novato, CA: New World Library, 2012).

53. Bronislaw Szerszynski, "Ecological Rites: Ritual Action in Environmental Protest Events," *Theory, Culture, Society* 19 (2002).

54. Fredericks, "Online Confessions of Eco-Guilt."

55. Conrad, "A Restorative Environmental Justice for Prison E-Waste Recycling."
56. Bell, *Ritual: Perspectives and Dimensions*, 145.
57. Joanna Macy and Molly Brown, *Coming Back to Life: The Updated Guide to the Work That Reconnects* (Gabriola Island, Canada: New Society Publishers, 2014); Joanna Macy and Chris Johnstone, *Active Hope: How to Face the Mess We're in without Going Crazy*.
58. I realize that I critique the existing ethical approaches for being overly conceptual and then take a fairly conceptual approach to my alternative. Yet I am proposing a reflective and action-oriented, communal response involving physical action and emotional and existential components that is significantly different from the narrowly rationalistic responses of many ethicists. I also recognize that implementing the kind of rituals I suggest here is a project that needs much more development beyond this chapter and will require collaboration with participants.
59. The sister city or synod programs that are popular ways to learn about other people and places and support each other through economic growth or spiritual support may be helpful models here. I would, however, argue for extending these models to a deeper recognition of past interactions, including harms, and future possibilities.
60. Of course, care is needed to avoid exploiting other participants, whether spiritually, materially, financially, or politically, especially if the resources or power between the two differed significantly.
61. For a discussion of similar issues, though with respect to local environmental degradation, see Jordan, *The Sunflower Forest*.
62. If geoengineering worked, it could give the initiator (or whoever claimed to be) a tremendous amount of power.
63. Certainly, this discussion has focused on SRM. Carbon capture rituals may necessitate distinctive features such as rituals to motivate and keep momentum and hope to complete massive distributed projects such as tree planting when their effects may yet be years away. Carbon capture initiatives may also need modes of envisioning and coordinating with people working all over the world for a common purpose.

BIBLIOGRAPHY

Bell, Catherine. *Ritual: Perspectives and Dimensions*. New York: Oxford University Press, 1997.

Bennett, Graceann, and Freya Williams. "Mainstream Green: Moving Sustainability from Niche to Normal." In *The Red Papers*. Chicago: Ogilvy & Mather, 2011.

Burkett, Maxine. "Climate Reparations." *Melbourne Journal of International Law* 10 (2009): 509–42.

Caney, Simon. "Environmental Degradation, Reparations, and the Moral Significance of History." *Journal of Social Philosophy* 37, no. 3 (2006): 464–82.

Carr, Wylie, Ashley Mercer, and Clare Palmer. "Public Concerns about the Ethics of Solar Radiation Management." In *Engineering the Climate: The Ethics of Solar Radiation Management*, edited by Christopher J. Preston, 169–86. Lanham, MD: Lexington Books, 2013.

Chavkin, Sasha. "Cash for Thunder: Bolivia Demands 'Climate Reparations.'" *Mother Jones*, 2009.

Clingerman, Forrest. "Between Babel and Pelagius." In *Engineering the Climate: The Ethics of Solar Radiation Management*, edited by Christopher J. Preston, 201–19. Lanham, MD: Lexington Books, 2013.

Clingerman, Forrest, and Kevin J. O'Brien. "Playing God: Why Religion Belongs in the Climate Engineering Debate." *Bulletin of Atomic Scientists* 70, no. 3 (2014): 27–37.

Cone Communications. "CSR Press Release: All That Glitters Isn't Gold: This Year, It's Green." (2007). Published electronically November 15, 2007. http://www.csrwire.com/press_releases/19356-All-That-Glitters-Isn-t-Gold-This-Year-It-s-Green.

Conrad, Sarah M. "A Restorative Environmental Justice for Prison E-Waste Recycling." *Peace Review* 23, no. 3 (2011): 348–55.

Corner, Adam, and Nick Pidgeon. "Geoengineering the Climate: The Social and Ethical Implications." *Environment* 51, no. 1 (2010): 24–37.

De Klerk, Barend J. "Enhancing Ecological Consciousness through Liturgical Acts of Doxology and Lament." *Verbum et Ecclesia* 35, no. 2 (2014): 1–8.

Eisner, Andrew. "Earth Day 2013 Special Gadgetology Report: Who Is Recycling Electronics and Buying Green Gadgets?" In Retrevo Blog. Los Angeles, CA: Retrevo, 2013. http://www.retrevo.com/content/blog/2013/04/earth-day-2013-special-gadgetology-report. Accessed January 17, 2016.

Ferguson, Mark A., and Nyla R. Branscombe. "Collective Guilt Mediates the Effect of Beliefs about Global Warming on Willingness to Engage in Mitigation Behavior." *Journal of Environmental Psychology* 30, no. 2 (2010): 135–42.

Figueroa, Robert Melchior, and Gordon Waitt. "Climb: Restorative Justice, Environmental Heritage, and the Moral Terrains of Uluru-Kata Tjuta National Park." *Environmental Philosophy* 7, no. 2 (2010): 135–63.

Fredericks, Sarah E. "Online Confessions of Eco-Guilt." *Journal for the Study of Religion, Nature and Culture* 8, no. 1 (2014): 64–84.

Freman, Kyle. "Survey: Does Eco Guilt Stop You from Shopping at Ikea?" Apartment Therapy, http://www.apartmenttherapy.com/survey-does-eco-64103.

Gardiner, Stephen M. *A Perfect Moral Storm: The Ethical Tragedy of Climate Change*. New York: Oxford University Press, 2011.

———. "Some Early Ethics of Geoengineering the Climate: A Commentary on the Values of the Royal Society Report." *Enviromental Values* 20 (2011): 163–88.

Hale, Ben. "The World That Would Have Been: Moral Hazard Arguments against Geoengineering." In *Engineering the Climate: The Ethics of Solar Radiation Management*, edited by Christopher J. Preston, 113–31. Lanham, MD: Lexington Books, 2013.

Hamilton, Clive. "The Ethical Foundations of Climate Engineering." http://clivehamilton.com/the-ethical-foundations-of-climate-engineering/.

Hourdequin, Marion. "Geoengineering, Solidarity, and Moral Risk." In *Engineering the Climate: The Ethics of Solar Radiation Management*, edited by Christopher J. Preston, 15–32. Lanham, MD: Lexington Books, 2013.

Jamieson, Dale. "Ethics and Intentional Climate Change." *Climatic Change* 33, no. 323–36 (1996).

———. *Reason in a Dark Time: Why the Struggle against Climate Change Failed—and What It Means for Our Future*. New York: Oxford University Press, 2014.

Jordan, William R. III. *The Sunflower Forest: Ecological Restoration and the New Communion with Nature*. Berkeley, CA: University of California Press, 2003.

Koch, Wendy. "Going Green, Women Show More Eco-Guilt in New Year's Resolutions." Web, *USA Today: Green House* (2009). Published electronically December 30, 2009. http://content.usatoday.com/communities/greenhouse/post/2009/12/going-green-women-show-more-eco-guilt-in-new-years-resolutions/1#.Ut1ifRDnaM-.

Kwenda, Stanley. "World: North Should Pay South Reparations for Climate Change." *Inter Press Service*, May 14, 2010.

Lexmark International, Inc. "Consumers Concerned, yet Clueless, on 'Green' Technology." Lexmark International, Inc. http://newsroom.lexmark.com/index.php?s=13630&item=23708.

Macy, Joanna, and Molly Brown. *Coming Back to Life: The Updated Guide to the Work That Reconnects*. Gabriola Island, Canada: New Society Publishers, 2014.

Macy, Joanna, and Chris Johnstone. *Active Hope: How to Face the Mess We're in without Going Crazy*. Novato, CA: New World Library, 2012.

Minteer, Ben. "Geoengineering and Ecological Ethics in the Anthropocene." *BioScience* 62, no. 10 (2012): 857–58.

Mooney, Maryellen, and Gina Stikes. "Ivillage and Today.Com 'Eco-Confessions' Survey, Most Parents Admit They Could Be Greener." http://www.nbcnews.com/id/47064299/ns/about-press_releases/#.Ut2KwxDnaM9.

National Geographic. "Greendex: Consumer Choice and the Environment—a Worldwide Tracking Survey." National Geographic, http://environment.nationalgeographic.com/envi-

ronment/greendex/?rptregcta=reg_free_np&rptregcampaign=20131016_rw_membership_r1p_us_se_w#.
National Geographic, and GlobeScan. "Greendex 2012: Consumer Choice and the Environment—a Worldwide Tracking Survey." 204. Washington, DC: National Geographic and GlobeScan, 2012.
Northcott, Michael S. "Buddhist Rituals, Mosque Sermons, and Marine Turtles: Religion, Ecology, and the Conservation of a Dinosaur in West Malaysia." *Journal for the Study of Religion, Nature & Culture* 6, no. 2 (2012): 196–214.
———. *A Moral Climate: The Ethics of Global Warming*. Maryknoll, NY: Orbis Books, 2007.
———. *A Political Theology of Climate Change*. Grand Rapids, MI: Wm. B. Eerdmans Publishing Co., 2013.
Pleasants, Jen. *Bag Green Guilt*. United States: Show the Love Media LLC, 2009.
Preston, Christopher J. "The Extraordinary Ethics of Solar Radiation Management." In *Engineering the Climate: The Ethics of Solar Radiation Management*, edited by Christopher J. Preston, 1–11. Lanham, MD: Lexington Books, 2013.
———. "Solar Radiation Management and Vulnerable Populations." In *Engineering the Climate: The Ethics of Solar Radiation Management*, edited by Christopher J. Preston, 77–93. Lanham, MD: Lexington Books, 2013.
Robock, A. "20 Reasons Why Geoengineering May Be a Bad Idea." *Bulletin of Atomic Scientists* 64, no. 2 (2008): 14–18.
Santmire, H. Paul. *Ritualizing Nature: Renewing Christian Liturgy in a Time of Crisis*. Minneapolis: Fortress Press, 2008.
Shelton Group. "American's Biggest 'Green' Guilt: Wasting Food, Survey Finds." Knoxville, TN, 2012.
Smart, Ninian. *Dimensions of the Sacred: An Anatomy of the World's Beliefs*. Berkeley: University of California Press, 1996.
Smith, Jonathan Z. *To Take Place: Toward Theory in Ritual*. Chicago Studies in the History of Judaism, edited by Jacob Neusner, William Scott Green, and Calvin Goldscheider. Chicago: The University of Chicago Press, 1987.
Szerszynski, Bronislaw. "Ecological Rites: Ritual Action in Environmental Protest Events." *Theory, Culture, Society* 19 (2002): 51–69.
Tangney, June Price, and Ronda L. Dearing. *Shame and Guilt*. New York: The Guilford Press, 2002.
The Royal Society. *Geoengineering the Climate*. London: The Royal Society, 2009.
van Wieren, Gretel. *Restored to Earth: Christianity, Environmental Ethics, and Ecological Restoration*. Georgetown: Georgetown University Press, 2013.
Vaze, Prashant. *The Economical Environmentalist: My Attempt to Live a Low-Carbon Life and What It Cost*. London: Earthscan, 2009.
von Jessen, Fritzie. *I Killed a Penguin: An Ecological Memoir*. Baltimore, MD: Publish America, 2009.

Chapter Ten

"First Be Reconciled": The Priority of Repentance in the Climate Engineering Debate

Kevin J. O'Brien

> When you are offering your gift at the altar, if you remember that your brother or sister has something against you, leave your gift before the altar and go; first be reconciled to your brother or sister, and then come and offer your gift.
>
> —Matthew 5:23–5, NRSV

This quote, attributed to Jesus, comes from the longest speech in the Christian New Testament, the Sermon on the Mount. In the midst of a series of moral instructions, Jesus instructs listeners to prioritize reconciliation over reverence, to "first be reconciled," and then to get about the business of worship. I find this a helpful way to begin an argument about climate engineering for two reasons. First, it reiterates a central premise of this book, that engineering the climate is a religious issue. This passage—and indeed the entirety of the Sermon on the Mount and much of the rest of the Bible—insists that religion in general and Judeo-Christian religions in particular cannot be compartmentalized to one day a week or to one part of life. Jesus's instructions are not merely about worship, but also how people treat their "brothers and sisters"—a group that, at least in one interpretation, includes every other person on earth. Thus, insofar as climate engineering is about our relationship to other people, it is relevant to religion.

Second, this text is about priorities, about getting the order of our actions right. One should not offer a gift to God until one has healed relationships with one's community. Reconciliation comes before worship. This discussion of priorities is a helpful model for discussions of climate engineering because the urgent question in the early twenty-first century is not "should we engineer the climate?" but instead "what do we need to do before deciding whether or not to engineer the climate?" While some climate engineering proposals depend on existing technologies, most will only be feasible with further research and development. Furthermore, no nations or political entities have yet considered climate engineering in an organized way, suggesting that there are important political and cultural questions that remain unresolved. At the time of this writing, people are considering and researching ways to engineer the climate, but are not ready to do so. This is therefore not the time to make final or absolute decisions, but rather to consider what will be necessary for such decisions.

Decisions about climate engineering will ultimately be political decisions, ideally made democratically. But scholars have a role in helping people to understand such decisions in an informed way. Wil Burns and Andrew Strauss make this point in *Climate Change and Geoengineering*, arguing that the job of a scholar is not to simply endorse or reject deliberate, large-scale manipulations of the climate, but rather to help inform a debate about "how we decide and who decides."[1] Again, this is about priorities: for scholars, the questions prior to climate engineering take precedence.

Perhaps the most important word in Burns and Strauss's question about "how we decide and who decides" is "we." Any discussion of climate engineering must include a very clear definition of who is assumed as the audience. The most commonly assumed "we" is the human species as a whole. Climate engineering would impact our entire species, and it is a proposed response to the global problem of climate change. However, as much as it would be ideal for the entire species to make a united decision, such global unity is distant, if not impossible. As philosopher Stephen Gardiner observes, humanity is not facing climate change as "one big, happy, global and intergenerational family." Instead, "climate impacts are spatially and temporally dispersed to different countries and generations, and in ways that affect the incentives of those involved to act."[2]

The "we" of this chapter includes people in the industrialized world who are relatively wealthy—those who disproportionately contribute climate-changing gases to the atmosphere, who can generally afford to avoid the most severe contemporary consequences of climate change, and who have options in our consumption and political structures that would allow us to reduce our impact on the climate. When I refer to "we," I refer to this group, in which I count myself. It is to this group that I will make my primary argument: our continued extensive contributions to climate change suggest that we cannot be trusted to make wise decisions about large-scale climate engineering. A moral approach to this issue in the industrialized world requires repentance as a first step. We must admit wrongdoing and cease to contribute to the problem before seeking to repair the damage we have already done.

The following section outlines an understanding of climate change, the problem that climate engineering attempts to solve, as structural violence. The next section then discusses the possibility of repentance and reparation as productive Christian responses to structural violence. Finally, the application of these ideas to climate engineering will be clarified by exploring the implications for two prominent positions on the issue. Each section contributes to the main argument: the only way privileged peoples can responsibly participate in moral discussions of climate engineering is to first repent for the ways we contribute to the problem we seek to solve.

NAMING THE PROBLEM

Climate engineering is a proposed response to anthropogenic climate change. Any perspective on climate engineering proposals therefore requires a definition of climate change; we must understand the problem before we can judge possible solutions. This is a difficult task, because climate change is a "wicked" problem, fundamentally difficult to understand and impossible to fully characterize from any one perspective.[3] Over the last two decades, as activists, politicians, and academics have developed perspectives and possible responses, it has become increasingly clear that there is no single definition of climate change—it is a scientific problem, an environmental problem, a human problem, an economic problem, a political problem, a religious problem, and so on. Each perspective is important, and none is sufficient by itself; each provides its own view of what is happening to the atmosphere, what the repercussions will be, and why it matters. Given this multiplicity of perspectives and the fact that changes to the atmosphere are irreversible, each perspective can offer *a* definition of climate change, but never *the* definition.[4] In that context, with no claim that any one view of climate change is adequate, I find it particularly helpful to understand it as a problem of structural violence.

In contrast to the "direct" violence of abuse, assault, or murder caused by an individual directly harming another, structural violence is harm perpetrated upon individuals or groups by social systems and ideologies, usually inflicted upon one group by another. Structural violence is indirect in the sense that no one person is responsible, no one person is at fault. This category has been widely used to characterize the problems of patriarchy, of racism, of classism; in each case, the violence has no single architect, and those that cause and benefit from it generally have the luxury of ignoring its existence. A white supremacist who actively seeks to harm people of color commits acts of direct violence. A white person who benefits from educational, economic, and political structures that make life easier for white people and harder for people of color participates in structural violence—usually without awareness or intention. Structural violence is subtle, at least for those who are not its victims. But it is nevertheless violence—people are hurt and killed by the actions of others. Human decisions create and sustain the structures that solidify and exacerbate power imbalances and hurt the victims of structural violence.[5]

To identify climate change as a problem of structural violence, then, is to observe that anthropogenic atmospheric changes cause and unjustly distribute pain and suffering. The commonly accepted forms of structural violence—classism, racism, sexism—are evident in and exacerbated by the forces that cause climate change. Perhaps the most obvious trends along these lines are economic: Refugees in poor nations and poor neighborhoods suffer as floods and droughts drive them from their homes. Residents of wealthy areas, by contrast, have so far been able to build seawalls, to import water, and to preserve their way of life in the face of a changing climate. Christopher Preston calls this the "moral deficit" of climate change, and importantly adds: "Taking this moral deficit seriously seems to be a prerequisite for a globally just and fair approach to geoengineering."[6]

Climate change also reflects the structural violence of racism and sexism. Communities of color tend to have the least access to the economic and political resources required to adapt to climate change, while Europeans and communities of European descent tend to have the most authority over and protection from existing institutions and their resources.[7] Women and children tend to bear the burdens of harsh agricultural conditions and food insecurity exacerbated by climate change.[8] Again, those who suffer the most from climate change have, in aggregate, contributed the least to the problem, while those who are most responsible for the problem (the "we" of this chapter) are the least likely to suffer its worst consequences.

To observe that climate change is *structural* violence signals a key challenge of taking these moral deficits seriously. Structural violence has no single architect, it is caused by no one person's intentions. Just as one can be entirely free from conscious sexism and yet still benefit from and perpetuate a patriarchal society, privileged people in the industrialized world cause climate change and benefit from our contributions to it without ever intentionally seeking to change the weather, raise the oceans, or destroy habitats. Literary scholar Rob Nixon captures this well by identifying climate change as a form of "slow violence," pain and harm that is "neither spectacular nor instantaneous, but rather incremental and accretive, its calamitous repercussions playing out across a range of temporal scales."[9] The waters on the coast of Bangladesh rise little by little, the wells of the Sudan dry up gradually, depriving those who live there of the essentials they need for survival. This is violence. But those of us far away have the luxury of ignoring these facts, ignoring the suffering to which we contribute. This is structural violence.

CLIMATE ENGINEERING AS A RESPONSE TO STRUCTURAL VIOLENCE

This understanding of climate change can helpfully inform discussions of climate engineering. In particular, it cautions us against the temptation to consider climate engineering as though it would occur in a vacuum, separate from historical power dynamics and the institutional structures that create and perpetuate climate change. When we recognize that climate change is structural violence, those of us who live in the privileged world and disproportionately cause the problem will recognize that we contribute to the causes of the problem climate engineering seeks to solve.

This perspective raises a critique of, for example, Steven Levitt and Stephen Dubner's dismissal of moral hazard arguments in *Superfreakonimics*. Many critics of climate engineering have raised the argument that the possibility of climate engineering creates a moral hazard, tempting political leaders and others to continue releasing climate-changing gases given the assurance that some future technology will prevent the harm it will cause.[10] Levitt and Dubner reject moral hazard arguments, writing that "blaming geoengineering" for the temptation to continue releasing climate-changing gases "is like blaming a heart surgeon for saving the life of someone who fails to exercise and eats too many french fries."[11] This analysis is incomplete, because it ignores the structural nature of climate change by separating those proposing climate engineering entirely from those who have caused the problem. If climate change is structural violence, the simile should be modified to include the fact that the doctor profits substantially from part ownership in a nearby french fry stand that advertises heavily and persuasively. Wealthy people in the industrialized world are complicit in changing the climate, and our discussions about whether to engineer it must take this fact seriously.

A full analysis of climate engineering must measure it as a response to structural violence, asking how it responds to harms caused by systems that are not readily apparent to those they benefit. Decisions about how we might manipulate the earth's atmosphere must include careful attention to how that manipulation will redress the unjustly distributed suffering of climate change.[12] Decisions about who might have the authority to intentionally manipulate the atmosphere must include careful attention to the fact that privileged people in the industrialized world have spent years making decisions that led to unintentional, harmful changes in the earth's climate. Understanding climate change as structural violence means that people of privilege must consider climate engineering as a solution to a problem we created, a way to redress a situation that has to this point disproportionately benefited us.

This does not predetermine an answer to the ultimate question of whether climate engineering could be a good idea. Advocates of climate engineering have emphasized the structural violence of climate change to stress the moral responsibility of wealthy people in the industrialized world, as when David Keith insists that the poor are "most likely to benefit from any short-term reduction in climate risk offered by geoengineering" and decries the "ugly prospect" of rich people rejecting climate engineering while failing to cut our emissions.[13] Climate engineering's critics also cite the injustice and invisibility of climate change; as when Clive Hamilton suggests that solar radiation management would "cement the failure of the North in its obligations to the global South," concluding, "If a just global warming solution cannot be found, who can believe in a just geoengineering regime?"[14] In both cases, climate change is understood as a problem of structural violence. The difference comes in whether climate engineering is understood to redress or exacerbate the problem.

My argument is that before we can answer that question—before we can decide whether climate engineering continues the injustices of climate change or begins to repair them—we need to face the structural violence in which we participate. Reconciliation is prior to any moral decision about climate engineering. Before trusting ourselves to make a clear judgment about what to do about a changing climate, privileged people in the industrialized world need to repent for changing the climate. This includes admitting the problem, lamenting its results, and converting away from the behaviors that have caused it.

THE IDEA AND PRACTICE OF REPENTANCE

Repentance is, of course, a religious word, and it will be explored here from a religious tradition. But repentance does not require particular religious beliefs: it simply means a decision to admit wrongdoing and to attempt to redress it. In Greek, the word for repentance, μετάνοια, means "thinking differently after." Repentance means changing one's mind, and therefore one's actions, after a transgression. To repent is to see that one has done something harmful and to seek to repair that harm. The Christian tradition offers tools that can inform any privileged person or community seeking to repent for structural violence.

South African theologian Ernst Conradie proposes that the world can learn from the South African response to apartheid, in which those guilty of profound direct and structural violence confessed their guilt in order to make a new society possible. South Africa's Truth and Reconciliation Commissions were by no means perfect, and the structural violence of racism is by no means abolished from that nation. However, new and healthier relationships were formed in part because many of the guilty recognized their failings, apologized for them, and shared a willingness to create a new and more just society. Religious leaders in South Africa led a process that helped make political and social healing possible. Conradie hopes that this can be repeated across the world in response to climate change: "Like the beneficiaries of apartheid it may be possible, by God's grace alone, to accept and specify one's responsibility towards causing climate change, to recognise one's guilt and show remorse."[15] Such confession and remorse may create a path to imagining new political, economic, and social systems that heal rather than degrade the global atmosphere. It is possible that engineering the climate will be an important part of that healing, but to decide that before genuinely repenting of climate change would be disordered, like seeking racial harmony without first facing the violence of apartheid.

The present essay takes its title from the biblical Gospel of Matthew, in which Jesus's first command is: "Repent, for the kingdom of heaven has come near" (4:17). One chapter later, he instructs believers to "leave your gift" at the altar and "first be reconciled" to their sisters and brothers. Written in the late first century for a community in the second generation of Jesus's followers, this gospel emphasizes the importance of community and the vital role that repentance plays in maintaining it. Communities are made up of fallible people who make mistakes, who hurt one another. Thus, healthy communities need systems and practices by which their members can turn away from wrongdoing, seek forgiveness from those who have been wronged, and redress harm. Repentance is a practice for keeping a community together, for moving on from mistakes and violence.

Based on Jesus's teaching that repentance is prior to reverence, many Christian churches begin each week's worship with a confession of failings, and rituals and disciplines have developed over millennia to foster repentance among Christians. Such repentance is not just about what happens in churches or within Christian communities; if it is meaningful, then it has broad social and political implications. Along these lines, Christian ethicist Jennifer McBride has developed a "theology of public witness," arguing that privileged Christian churches should define themselves by their willingness to admit, face, and turn away from wrongdoing in political life as well as worship. She identifies the Christian community as "the body called to repentance—as the body called . . . to accept guilt or take responsibility for social sin."[16]

Violence and oppression separate people not only from one another but also from God, and so privileged people can only move closer to the divine by distancing ourselves from the violence we cause—unintentionally as well as intentionally. Christian repentance involves a decision to side with God against the social structures and systems that harm others. This theological perspective assumes that a Christian's relationship to God is defined by her relationship to other persons: to repent for harming one's neighbor is to restore our relationship to the divine. This helps to explain why reconciliation with one's "brother or sister" precedes worship.

REPENTING FOR STRUCTURAL SIN

Brazilian liberation theologian Gustavo Gutiérrez also argues that the Biblical call to "first be reconciled" applies to political structures. He emphasizes that this teaching about repentance is in line with the Hebrew Bible's prophetic tradition's teaching that "worship is authentic only when it is based on profound personal dispositions, on the creation of true fellowship, and on real commitment to others, especially the most needy."[17] Writing about the structural violence of poverty and political injustice in Latin America, Gutiérrez suggests that contemporary Christians are called to "continual conversion," a willingness to "see and admit what is vitiating our relationship with God and our solidarity with others—what, in consequence, is also hindering the creation of a just and human society." Such conversion includes both internal soul searching and a willingness to reshape "the web of social relationships of which the individual is a part."[18]

Applying such thinking to racism in the United States, Christian ethicist Jennifer Harvey criticizes whites who seek equality and reconciliation between races without first repenting of and seeking to repair the harms caused by structural violence. Racism creates enduring "legacies of unaddressed violence, oppression, subjugation, and devastation for which those of us who have benefited have yet to apologize, let alone make meaningful repair."[19] In an argument resonant with Matthew 5:23, Harvey suggests that white Christians have no right to seek worshipful reconciliation until they offer meaningful, material repentance. In the face of structural violence, Harvey insists, the privileged cannot be reconciled with the oppressed until we have apologized for and made concrete attempts to redress the damage we have caused.

These authors write from very different contexts, but both insist that Christians are called to repent of structural violence. Reading them side-by-side reveals three distinct steps for such repentance: recognizing the harm we have caused, apologizing for it, and then working to redress the wrong. The first step is recognition, becoming aware of structural violence and the harms it does to our neighbors. For Gutiérrez, this means recognizing the political and economic colonialism that causes oppression in Latin America while primarily benefiting wealthy persons in Europe and North America. He argues that anyone seeking to live a Christian life must engage in "concrete analysis" of this imbalance, quoting Oscar Romero: "Nowadays an authentic Christian conversion must lead to an unmasking of the social mechanisms that turn the worker and the peasant into marginalized persons."[20]

The second step is apology, expressing awareness that one is not just a bystander to structural violence but also a perpetrator, and developing a plan to stop causing this harm. Along these lines, Harvey suggests that before any steps toward a "beloved community" of racial harmony can commence, "white Christians need first to issue an apology for ongoing complicity in white supremacist structures that continue to harm African Amcricans." She stipulates that such an apology must include clear statements of "what is being apologized for" and "what white Christians intend to do to rectify the harm and actively refuse to oppress again."[21]

Only after the violence has been recognized and its perpetrators have apologized is it reasonable to take the third step, redressing the wrong and seeking to make a world that is less violent. Gutiérrez imagines the liberation of Latin America from oppression, and Harvey imagines a future of racial reconciliation, but both caution against jumping to those endpoints without carefully analyzing the extent of the problem and our own complicity in it. Thus, Gutiérrez suggests that only those who aligned themselves to the poor and oppressed by meaningfully surrendering their privilege can begin to define what liberation would look like. Harvey insists that whites cannot be trustworthy interlocutors in discussions about race until we have demonstrated our seriousness by admitting our debt and offering reparations.

This three-step process emphasizes how much is involved in putting reconciliation first, in repenting meaningfully for structural violence before taking further action. This cautious approach, recognizing and limiting harm before believing one can redress it, has much to teach privileged persons concerned about our complicity in climate change.

REPENTING FOR CLIMATE CHANGE

To repent of climate change means, first, to acknowledge the extent of the problem. The analysis of climate change as structural violence above seeks to work toward this goal, exploring the depth of harm caused by a changing climate. Ethicist Cynthia Moe-Lobeda carries this work further, beginning her analysis of climate change by "unmasking evil." She argues that a key power of structural violence is its "relative invisibility" to those who do not suffer from it, and so carefully develops an account of how dominant economic, political, and social structures harm the poor and the environment. While such systems are often masked in a disguise of inevitability—the market cannot be controlled, the poor are poor by their own choices, human development will inevitably change the climate—she argues that structural violence must always be recognized as a human creation. Ultimately, she insists that this kind of analysis is empowering: the political and economic structures that are changing the climate "were constructed by human beings and therefore can be changed by them. The neoliberal global economy—including its manifestation in national economies—was constructed by people. It can therefore be replaced."[22] To unmask structural sin is to take the first step of resisting it.

The second step of repentance, apology for our own complicity, involves facing truths that are very personally disturbing, recognizing the ways our choices are harming other people—indirectly, but still with dire consequences. When I drive a car or drink milk produced in a factory farm, I am contributing to the release of CO_2 and methane into the atmosphere, and so I am contributing to the flooding in the Maldives and the increase in average hurricane strength across the world. When energy executives prioritize their investments in fossil fuels and the continued economic health of their corporations, they may intend to help their investors and employees, but they are nevertheless making decisions that degrade ecosystems and human lives. When politicians refuse to pass laws that restrict climate-changing emissions, they may be responding to the short-term needs and desires of their constituents and campaigns, but they are nevertheless causing long-term harm to the planet's ecosystems. No one intends to cause climate change, but we are all part of the problem. Repentance requires that we face this fact and apologize for the damage we are causing.

For some, awareness leads not to apology, but paralysis—feelings of complicity, guilty, or hypocrisy lead to despair rather than action. So, Willis Jenkins has suggested that people need practices of confession, traditions that structure and inform meaningful and substantive apology. He argues that such practices may in fact be one of the most important contributions Christians can make to ongoing discussions of climate change.[23] Along similar lines, Sarah Fredericks demonstrates in the previous chapter of this volume that rituals offer a productive means to engage a problem as complicated as climate change and thereby intervene in debates about climate engineering. Repentance requires avoiding paralysis as we apologize and take responsibility for the harms in which we participate, recognizing our role in the violence of climate change in such a way that we can act on our recognition.

Only after the steps of recognition and apology have been taken can the process of repentance for climate change move to genuinely redress the problem. This step is the furthest from current reality, and so the most abstract. Perhaps it would involve reparations for the damage of climate change, or reorganized global economic and political structures. It might also involve climate engineering, which is reasonable to consider as a way for privileged people to undo the violence currently inflicted disproportionately on the poor, people of color, women, and children. However, it is impossible to defend that claim as yet: one cannot anticipate the realization of repentance when one has not yet begun the process of recognition and apology. Thus, even if climate engineering is a part of redressing the violence of climate change, we could only legitimately undertake it after a genuine accounting of the problem and our role in it.

HOW REPENTANCE CHANGES THE CONVERSATION ABOUT CLIMATE ENGINEERING

Having defined repentance as a response to the structural violence of climate change and demonstrated the resources Christian ethics offers to such a response, I turn now to make the argument slightly more concrete. To demonstrate my claim that repentance must be prior to a decision about climate engineering, I will engage one advocate of climate engineering—writer and activist Jay Michaelson—and one opponent—philosopher Clive Hamilton.

The Master's Tools: We Should Not Accept Climate Engineering before Repenting

In what I see as one of the strongest arguments made in favor of climate engineering, Jay Michaelson suggests that what he calls "climate management" is "*necessary* because of the unfortunate political economy of [greenhouse gas] reduction, in which the highest costs of mitigation fall precisely on some of the most powerful political actors." Climate management is "inevitable," Michaelson argues, because "it is the lowest-cost option that appeals to the widest range of political actors."[24]

Identifying himself as a liberal environmentalist, Michaelson admits that he finds the idea of working with corporate interests to manage the climate "nauseating." He explains: "First, greedy oilmen block [greenhouse gas] reduction by saying that climate change does not exist, and then they support [climate engineering] because all the sudden it does." But, Michaelson argues, this willingness of corporate interests to engage climate change is ultimately "a blessing, not a curse," because it gives climate engineering a political chance far greater than any exclusively grassroots, progressive proposal. He asks: "Are we environmentalists interested in punishing the bad guys, or saving the earth?"[25]

Michaelson's argument heavily emphasizes the need to act on climate change, and so understandably advocates compromise in service of progress. My argument is less consequentialist and more focused on virtue ethics; rather than emphasizing that action must be taken, I am interested in who the actor is.[26] While I agree with Michaelson that our response to climate change should not be about "punishing the bad guys," it is nevertheless vital to properly identify who the bad actors are. As an instance of structural violence, climate change is not caused exclusively by a few "greedy oilmen." All of us who regularly buy and use products made from petroleum, consume electricity from coal-fired power plants, eat food grown on industrial farms, and travel in airplanes and cars share complicity. Those of us who indulge in these behaviors more than is necessary to meet our basic needs have disproportionate complicity. Because people of privilege are not yet capable of recognizing this fact as a group, we are not trustworthy to make wise decisions about climate management.

Michaelson thoughtfully quotes Audre Lorde's famous adage that "the master's tools will never dismantle the master's house," but admits that he is advocating "playing by the master's rules—the political arithmetic determined by financial power." He continues: "If environmentalists are serious about climate change, then we must divest ourselves of our utopian dreams and dystopian nightmares and do the messy work of political compromise."[27] I agree that pure idealism is problematic, but am less willing to dismiss Lorde's principle. Messy political compromise may be necessary, but it is only worthwhile if it moves in a constructive direction, and we will only recognize what is constructive if we can clearly identify what has been destructive in the past. Lorde asks us to carefully examine existing structural violence and the mechanisms by which it perpetuates itself. Until we have done that, until we have examined the master's tools, we are unequipped to make responsible decisions about whether or not to use them.

"The master's house" is the system in which we live, which includes substantial structural violence; "the master's tools" are the institutions and ideologies that perpetuate that system. We are stuck in a violent system and the structures of that system seem to be the only ones available, but until we recognize these facts, we will not know what our options are. This awareness of structural forces is more constructive than Michaelson's discussion of environmentalist "good guys" who work against the "bad guys" in the oil industry who destroyed the climate. We do not live in a world of good guys and bad guys, and so the question is not whether "we" should compromise with "them" in order to engineer the climate. Instead, we live in a world of complex institutions and ideologies, and we must decide how and whether to use the tools of that system after determining whose interests we are serving in doing so.

It is not a foregone conclusion that climate engineering would be "the master's tools," inevitably replicating existing injustices. But until we acknowledge that we are living in the master's house, that our perspectives on climate change and climate engineering are shaped by our privilege, we must consider the possibility. So, we are not ready to accept climate engineering.

Avoiding Moral Failure: We Should Not Reject Climate Engineering before Repenting

In contrast to Michaelson, the philosopher Clive Hamilton makes an argument against climate engineering that is heavily informed by awareness of privilege and the structures that have caused climate change and unjustly distributed its burdens. Hamilton reminds readers that "consideration of climate engineering comes after a long history of bad faith in international negotiations, in which various nations have engaged in pious declarations of concern coupled with ruthless obstructionism." He argues that intentionally managing the climate on a large scale would simply be a continuation of this "moral failure," and so all proposals for climate engineering should be viewed with profound skepticism.[28]

Hamilton's is an argument of virtue ethics: he contrasts the mechanistic faith in technology required to engineer the climate with the more laudable virtue of humility, accepting human limitations "in the face of the superior power, complexity, and enigmatic character of the earth."[29] The contrast is clear: those who seek to master the Earth have a dangerously inflated faith in human power, those who resist climate engineering are appropriately and virtuously humble.

Insofar as Hamilton argues that wealthy people in the industrialized world do not currently have moral ground from which to advocate climate engineering, I agree completely. However, at times he goes further than this to suggest that any management of the climate would always and inevitably be a mistake, that any attempt at climate engineering is inherently an expression of hubris, treating the earth as an object to be subdued.[30] I am less certain about this. One reason is suggested by Hamilton himself: should oppressed peoples from poorer nations advocate climate engineering, their moral claims would need to be adjudicated on entirely different moral terms. While I know of no enthusiastic and sustained call for climate engineering from poor and oppressed peoples, such a call would need to be taken very seriously on its own terms. If those who have *not* contributed to climate change and who are therefore *not* complicit in its structural violence seek to engineer the climate—and demand assistance in doing so from the industrialized world—it would be unfair to assume they were expressing hubris. Calling the perpetrators of climate change to repentance and humility involves genuinely listening to others, and if others seek to engineer the climate, those of us who caused the problem must listen openly.

I would go further and suggest that climate engineering need not necessarily be an expression of hubris because I believe in the possibility of conversion. If we believe that moral transformation is possible, then we must not prejudge the virtue of people who might someday achieve moral awareness beyond our current complicity. The process of repentance insists upon a clear awareness of wrongdoing, a meaningful apology for it, and then an attempt to redress the damage done. Wealthy communities in the industrialized world have yet to begin this process in a meaningful way, have yet to fully realize the extent of climate change's structural violence, or apologize for our role in causing it. Engineering the climate could, potentially, be a meaningful expression of repentance, redressing the harm future people of privilege have recognized and apologized for. It is inappropriate to say whether or not such activity would be morally legitimate—or even necessary—until after a genuine admission of guilt and a true cessation of the processes contributing to climate change. We have not begun the process of repentance, and so we are not trustworthy to determine how it should end.

Hamilton suggests that climate engineering would necessarily continue the "moral failure" of hubristically changing the climate and forcing others to bear the consequences. I am less certain that we privileged peoples yet understand our moral failure. There is more work to do truly admitting our mistakes and turning away from them. I am therefore less certain than Hamilton that I know what a meaningful redress would look like. Perhaps climate engineering will be the best we can do to repair the damage we have caused; we will not know until we have honestly assessed it, after honestly assessing ourselves.

CONCLUSION: THE VIRTUE OF PATIENCE

It is important to speak about climate engineering from a particular social location, because this allows us to first discern what we have the right to say and what we do not. I have argued that "we" privileged people in the industrialized world do not have the right to make a blanket statement about whether climate engineering could be acceptable. Prior to answering that question, we must repent for our role in making the question necessary. Repentance requires an admission of complicity and cessation of the behaviors causing the problem before assuming we know how to redress the harm. The definition of repentance as "thinking differently after" implies a need for patience: we cannot think as well as we need to until we have repented.

Moral discussions of climate change are frequently tinged with urgency—human populations are already suffering, change is already inevitable, and important decisions must be made. This is all true; the problems facing humanity and the ecosystems upon which humanity depends are immediate and enormous. However, facing an immediate and enormous problem does not necessarily make one equipped to offer a wise and prudent response, and action is not always better than inaction.

In chapter 2 of this volume, Thomas Bruhn, Stefan Schäfer, and Mark Lawrence report that their transdisciplinary conversation with Christians about climate engineering called for "Solominic wisdom" on this difficult topic. Such wisdom requires patience; wisdom is about the depth of one's decisions, not their speed. A wise response to climate change will come only after careful and deliberate consideration. The problems might be immediate, but wise responses will take time.

As mentioned in the introduction, few advocates of climate engineering believe that we should implement it immediately, and most call for study rather than abrupt action. Consider David Keith: "The question at hand is not immediate implementation, but whether or not to begin a serious research program that has strong mechanisms for independent review and is coupled with the development of governance mechanisms that could enable legitimate decisions about larger scale research and implementation."[31] There are dangers to studying climate engineering, well enumerated by its critics,[32] but Keith makes a reasonable argument that it is worthwhile to study what is possible as long as such study is open to suspicion and critique.

I would also advocate a program of rigorous and widespread *moral* study—we need to better understand and debate our responsibilities and the rights of other peoples, future generations, and nonhuman animals. Such study should lead to moral action—those of us complicit in the structural violence of climate change need to genuinely admit that complicity, apologize for it, and cease the damage we are doing before believing we are equipped to repair it. If our moral studies lead to different results than our technical studies—teaching us that we should not do things that we are capable of, for instance, then that will be something worth talking about.[33]

I suggest, in other words, that any "serious research program" on climate engineering must include careful studies of human virtue. No discussion of climate engineering among privileged peoples is complete unless it raises the question of what could make us trustworthy to make decisions about managing the atmosphere in collaboration with others. Christians have something to contribute to this conversation, partly through our study and practice of repentance. Others will have other contributions, which will hopefully carry forward a dialogue about climate change and climate engineering that seeks to reconcile human communities with one another and with the natural systems upon which we depend.

NOTES

1. "Introduction" in William C. G. Burns and Andrew L. Strauss, *Climate Change Geoengineering: Philosophical Perspectives, Legal Issues, and Governance Frameworks* (New York: Cambridge, 2013), 2.

2. Stephen Mark Gardiner, "Some Early Ethics of Geoengineering the Climate: A Commentary on the Values of the Royal Society Report," *Environmental Values* 20 (2011): 172.

3. For the original discussion of wicked problems, which focused on urban planning, see Horst J. W. Rittel and Melvin M. Webber, "Dilemmas in a General Theory of Planning," *Policy Sciences* 4, no. 2 (1973).

4. In a way that has been influential upon and is hopefully compatible with the approach of this chapter, Willis Jenkins argues that understanding climate change as a wicked problem justifies a pragmatic approach to contemporary Christian ecological ethics: Willis Jenkins, *The Future of Ethics: Sustainability, Social Justice, and Religious Creativity* (Washington, DC: Georgetown University Press, 2013), ch 4.

5. See Cynthia D. Moe-Lobeda, *Resisting Structural Evil: Love as Ecological-Economic Vocation* (Minneapolis, MN: Fortress Press, 2013), 71–78.

6. Christopher J. Preston, "Solar Radiation Management and Vulnerable Populations: The Moral Deficit and Its Prospects," in Christopher J. Preston, ed., *Engineering the Climate: The Ethics of Solar Radiation Management* (Lanham, MD: Lexington Books, 2012), 79.

7. See, for example, Congressional Black Caucus Foundation, "African Americans and Climate Change: An Unequal Burden" (2004), http://rprogress.org/publications/2004/CBCF_REPORT_F.pdf (accessed August 21, 2015).

8. See, for example, UN WomenWatch, "Women, Gender Equality and Climate Change," http://www.un.org/womenwatch/feature/climate_change/ (accessed August 21, 2015). It is beyond the scope of the present chapter, but I argue elsewhere that climate change should also be understood as a form of structural violence perpetrated against future generations and nonhuman animals. See Kevin J. O'Brien, *The Violence of Climate Change* (Georgetown University Press, forthcoming).

9. Rob Nixon, *Slow Violence and the Environmentalism of the Poor* (Cambridge, MA: Harvard University Press, 2011), 2.

10. For a review of moral hazard arguments, and then a stronger case against their validity, see especially Ben Hale, "The World That Would Have Been: Moral Hazard Arguments against Geoengineering," in Preston, *Engineering the Climate*.

11. Steven D. Levitt and Stephen J. Dubner, *Superfreakonomics: Global Cooling, Patriotic Prostitutes, and Why Suicide Bombers Should Buy Life Insurance* (New York: William Morrow, 2009), 197. Levitt and Dubner attribute the simile to Nathan Myhrvold.

12. This is argued well, with a particular focus on the structural injustice of patriarchy, in Holly Jean Buck, Andrea R. Gammon, and Christopher J. Preston, "Gender and Geoengineering," *Hypatia* 29, no. 3 (2014).

13. David W. Keith, *A Case for Climate Engineering* (Cambridge, MA: MIT Press, 2013), 137.

14. Clive Hamilton, *Earthmasters: The Dawn of the Age of Climate Engineering* (New Haven, CT: Yale University Press, 2014), 163.

15. Ernst M. Conradie, *The Church and Climate Change* (Pietermaitzburg, South Africa: Cluster Publications, 2008), 88. Cynthia Moe-Lobeda similarly hopes that the churches will make genuine repentance possible, asking: "Could it be that worship that empowers the people of God for social and ecological healing will include profound lament for the ways in which our lives unwittingly endanger Earth's life-systems and vulnerable neighbors far and near?" (*Resisting Structural Evil*, 262).

16. Jennifer M. McBride, *The Church for the World: A Theology of Public Witness* (New York: Oxford University Press, 2012), 9.

17. Gustavo Gutiérrez, *A Theology of Liberation: History, Politics, and Salvation* (Maryknoll, NY: Orbis Books, 1988), 132.

18. Gustavo Gutiérrez, *We Drink from Our Own Wells: The Spiritual Journey of a People* (Maryknoll, NY: Orbis Books, 2003), 97–98.

19. Jennifer Harvey, *Dear White Christians: For Those Still Longing for Racial Reconciliation* (Grand Rapids, MI: Eerdmans, 2014), 5.

20. Gutiérrez, *We Drink from Our Own Wells*, 98.

21. Harvey, *Dear White Christians*, 98.

22. Moe-Lobeda, *Resisting Structural Evil*, 230.

23. Jenkins, *The Future of Ethics*, 57.

24. Jay Michaelson, "Geoengineering and Climate Management: From Marginality to Inevitability," in Burns and Strauss, *Climate Change Geoengineering*, 81, 107. Emphasis in original.

25. Ibid., 82.

26. I thank Toby Svobada for his thoughtful analysis of virtue approaches to this issue in "Climate Engineering and Virtuous Agents," presented at the *Seizing an Alternative* Conference, Claremont, CA, June 5, 2015.

27. Michaelson, "Geoengineering and Climate Management," 110.

28. Hamilton, *Earthmasters*, 162–63, 165–66.

29. Clive Hamilton, "The Ethical Foundations of Climate Engineering," in Burns and Strauss, *Climate Change Geoengineering*, 48.

30. He writes: "Climate engineering represents a conscious attempt to overcome resistance of the natural world to human domination, the last great stride toward total ascendancy" (ibid., 56).

31. Keith, *A Case for Climate Engineering*, 149.

32. See, for example, Gardiner, "Some Early Ethics of Geoengineering."

33. Any interdisciplinary project should be open to such conflicts, to the possibility that one perspective disagrees with another and decisions must be made. On this see especially Bronislaw Szerszynski and Maialen Galarraga, "Geoengineering Knowledge: Interdisciplinarity and the Shaping of Climate Engineering Research," *Environment and Planning* 45, 12 (2013).

BIBLIOGRAPHY

Buck, Holly Jean, Andrea R. Gammon, and Christopher J. Preston. "Gender and Geoengineering." *Hypatia* 29, 3 (2014): 651–69.

Burns, William C. G., and Andrew L. Strauss. *Climate Change Geoengineering: Philosophical Perspectives, Legal Issues, and Governance Frameworks*. New York: Cambridge, 2013.

Conradie, Ernst M. *The Church and Climate Change*. Pietermaitzburg, South Africa: Cluster Publications, 2008.

Gardiner, Stephen Mark. "Some Early Ethics of Geoengineering the Climate: A Commentary on the Values of the Royal Society Report." *Environmental Values* 20 (2011): 163–88.

Gutiérrez, Gustavo. *A Theology of Liberation: History, Politics, and Salvation*. Maryknoll, NY: Orbis Books, 1988.

———. *We Drink from Our Own Wells: The Spiritual Journey of a People*. Maryknoll, N.Y: Orbis Books, 2003.

Hamilton, Clive. *Earthmasters: The Dawn of the Age of Climate Engineering*. New Haven, CT: Yale University Press, 2014.

Harvey, Jennifer. *Dear White Christians: For Those Still Longing for Racial Reconciliation*. Grand Rapids, MI: Eerdmans, 2014.

Jenkins, Willis. *The Future of Ethics: Sustainability, Social Justice, and Religious Creativity*. Washington, DC: Georgetown University Press, 2013.

Keith, David W. *A Case for Climate Engineering*. Cambridge, MA: MIT Press, 2013.

Levitt, Steven D., and Stephen J. Dubner. *Superfreakonomics: Global Cooling, Patriotic Prostitutes, and Why Suicide Bombers Should Buy Life Insurance*. New York: William Morrow, 2009.

McBride, Jennifer M. *The Church for the World: A Theology of Public Witness*. New York: Oxford University Press, 2012.

Moe-Lobeda, Cynthia D. *Resisting Structural Evil: Love as Ecological-Economic Vocation*. Minneapolis: Fortress Press, 2013.

Nixon, Rob. *Slow Violence and the Environmentalism of the Poor*. Cambridge, MA: Harvard University Press, 2011.

Preston, Christopher J., ed. *Engineering the Climate: The Ethics of Solar Radiation Management*. Lanham, MD: Lexington Books, 2012.

Rittel, Horst J. W., and Melvin M. Webber. "Dilemmas in a General Theory of Planning." *Policy Sciences* 4, 2 (1973): 155–69.

Szerszynski, Bronislaw, and Maialen Galarraga. "Geoengineering Knowledge: Interdisciplinarity and the Shaping of Climate Engineering Research." *Environment and Planning* 45, 12 (2013): 2817–24.

Appendix

Religion and Climate Engineering: Points of Consensus from Claremont

Thomas Bruhn, Forrest Clingerman,
Sarah Fredericks, Laura Hartman, Kevin J. O'Brien,
Dane Scott, and Marit Trelstad

Scholars have only just begun to consider the intersection of religion and climate engineering. There are still uncertainties over a great many things, such as the methods with which to approach the topic, the role of religious studies in confronting society's problems, and the relationship between climate science and humanistic reflection. More pressingly, as of yet there is no general agreement on the role of religious communities and scholars of religion in creating space for moral reflection on climate engineering.

In the context of the *Seizing an Alternative: Towards an Ecological Civilization* conference held in Claremont, California, on June 4–7, 2015, a group of scientists, theologians, and ethicists met to share scholarly reflections on climate engineering. A core group met as a seminar for three days. Session time was focused on discussion and debate, centered on the early drafts of the essays that became this book. Such discussions naturally led to questions over what role theology and ethics have in the debate, as well as the tools religion offers for moral discussions.

After the sessions at Claremont, Thomas Bruhn and Marit Trelstad asked participants to summarize what they took to be the consensus that emerged from the discussion. Bruhn and Trelstad compiled the initial list in conversation with conference participants during the summer of 2015. Through a collaborative process of editing, this list was refined in January 2016. These points summarize areas of agreement that emerged during conversations at Claremont and reported by participants. While preliminary, these points of consensus will hopefully provide a starting point for further discussions on religion and climate engineering.

THE CLAREMONT POINTS OF CONSENSUS

- The debate over climate engineering urges humans to critically reflect on our self-understanding, especially on the role of humankind on earth and the conditions under which humans can or may intervene in ecosystems. In this pursuit of self-understanding, religious groups have a particularly important perspective to contribute. It therefore seems reasonable for those interested in or concerned about climate engineering to seek a proactive engagement with religious groups.
- Religious communities and moral traditions have a role in the public sphere. This role includes participating in debates over moral questions, critiquing policy decisions from normative perspective, and suggesting narrative frameworks for understanding.
- At the same time, individual religious communities should recognize the religious diversity of the global public square. Political and social assessments of climate engineering cannot use a single religion as the sole normative framework for global discussions.
- Ethical arguments can be made for and against the idea of climate engineering. Because of this moral ambiguity, climate engineering should be neither rejected nor endorsed prior to serious moral engagement and reflection.
- Whether or not climate engineering proposals go forward, emissions reduction is essential.
- Climate engineering must not focus on covering up or enabling environmental damage. If pursued at all, it will ideally address both the impacts and the root causes of climate change.
- Discussions of climate engineering, like discussions of all responses to climate change, must take seriously the differences in power and privilege between rich and poor—as well as developed and undeveloped—peoples and nations. This means that national and international discussions should work to include people representing diverse economic, racial, social, national, cultural, and religious backgrounds. Public conversations on this topic should include laypeople and citizens of all types. Furthermore, all climate engineering proposals should include assessment of potential impacts on various communities, both human and nonhuman.
- Conceptual models, metaphors, and stories are often helpful for engaging complicated topics; they have a large role in framing the issues surrounding climate engineering. Therefore, it is important to carefully consider

our language and how it impacts our understanding of the problem as well as potential responses.

- Acknowledging our own personal and social culpability for global climate change, conceptually and/or ritually, may be necessary before we can claim real authority to act or to make good decisions. Religious practices of repentance can help people to understand the power of accountability and call them into new action. Repentance, however, should not be used to justify harmful actions, avoid the issue of climate change, or to avoid making decisions about climate engineering.

Contributors

Thomas Bruhn received a doctorate in physics from the Technical University of Berlin for his dissertation about the interaction between organic molecules and semiconductor nanostructures. After working for the Leibniz institute for analytical sciences (ISAS) and research stays at University "Tor Vergata" in Rome (Italy) and the Interdisciplinary Centre for Nanosciences (CINaM) in Marseille (France), he joined the Institute for Advanced Sustainability Studies (IASS) Potsdam in 2012. His research focuses on environmental impacts of carbon capture technologies and stakeholder engagement about climate engineering. Beyond this, he is particularly interested in the changing role of humankind in the earth system in the light of the Anthropocene and how the cultivation of certain mind-sets among humans can contribute to the societal transformation to sustainability. Beside his research, he is active in several NGOs. He is a member to the thinktank30 of the Club of Rome and a member of the board of directors of the Global Contract Foundation.

Forrest Clingerman is associate professor of religion and philosophy at Ohio Northern University. He is co-editor of *Placing Nature on the Borders of Religion, Philosophy and Ethics* (Ashgate, 2011) and *Interpreting Nature: The Emerging Field of Environmental Hermeneutics* (Fordham University Press, 2013). In addition he has published a number of journal articles and book chapters on environmental thought. His publications have focused on the interpretation of place; the relationship between nature and the arts; and theological responses to climate change and the Anthropocene.

Sarah E. Fredericks is assistant professor of environmental ethics at the University of Chicago Divinity School. She studies the interrelationship of science, philosophy, and religion in environmental ethics. She is particularly interested in sustainability, energy, environmental guilt and shame, the ethics of international environmental policy-making, and environmental justice. Her current book project examines how collective guilt and shame about our contributions to climate change are ethical issues because they can exacerbate our contributions to climate change or help us take action to slow climate change. She argues that religious rituals as modes of action, identity-formation, and reconciliation are necessary responses to moral emotions about climate change. Fredericks is the author of *Measuring and Evaluating Sustainability: Ethics in Sustainability Indexes* (Routledge, 2013) and articles in the *Journal for the Study of Religion, Nature and Culture*; *International Journal of Sustainable Development and World Ecology*; *Environmental Justice*; and *Ethics, Policy, and Environment*; as well as several book chapters.

Laura M. Hartman is assistant professor of environmental studies at the University of Wisconsin-Oshkosh. Her book, *The Christian Consumer: Living Faithfully in a Fragile World* was published by Oxford University Press in 2011. Her forthcoming edited volume, *Flourishing: Comparative Religious Environmental Ethics*, will also be published by Oxford. She has published articles about environmental virtue ethics, Sabbath keeping, food justice, transportation ethics, and geoengineering.

Willis Jenkins is associate professor of religion, ethics, and environment at the University of Virginia. He is the author of two award-winning books: *Ecologies of Grace*: *Environmental Ethics and Christian Theology* (Oxford 2008), which won a Templeton Award for Theological Promise, and *The Future of Ethics: Sustainability, Social Justice, and Religious Creativity* (Georgetown 2013), which won an American Academy of Religion Award for Excellence in the Study of Religion. He is also editor or co-editor of several books, including the forthcoming *Routledge Handbook to Religion and Ecology*.

Mark G. Lawrence is scientific director at the Institute for Advanced Sustainability Studies (IASS) in Potsdam, focusing on a range of sustainable development topics at the science-policy and science-society interface, including mitigating the impacts of short-lived, climate-forcing pollutants (SLCPs), and on the potential impacts, uncertainties, and risks of "climate engineering." He received his Ph.D. in 1996 from the Georgia Institute of Technology. From 2000 until 2011 he was a research group leader at the Max Planck Institute for Chemistry in Mainz. He served as interim professor for meteorology at the University of Mainz during 2009–2010, and moved to the IASS in 2011. In 2014 he was appointed Honorary Professor at the University of Potsdam. Prof. Lawrence is author or co-author of over one hundred peer-reviewed publications. He has led various international projects, has served as editor for two journals (*Atmospheric Chemistry and Physics*; *Atmospheric Environment*), and is on several international committees, including being co-chair of the International Global Atmospheric Chemistry program (IGAC).

Kevin J. O'Brien is associate professor of religion studies at Pacific Lutheran University. His teaching and research focus on Christian environmental ethics, the intersection of religion and the environment, and the intersection of environmental concerns with social justice. He wrote *An Ethics of Biodiversity* (Georgetown, 2010), co-wrote *An Introduction to Christian Environmentalism* (Baylor, 2014), and co-edited *Grounding Religion* (Routledge, 2010) and *Inherited Land* (Cascade, 2011). He is currently writing a book about the relevance of the Christian nonviolent tradition to the contemporary movement for climate justice.

Stefan Schäfer leads the research program "Emerging Technologies and Social Transformations in the Anthropocene" at the Institute for Advanced Sustainability Studies in Potsdam, Germany. His research focuses on the history, philosophy, and politics of science and technology, with a particular focus on the emerging field of climate engineering. He was a guest researcher at the Social Science Research Center Berlin (WZB) from 2009–2012 and a fellow of the Robert Bosch Foundation's Global Governance Futures program in 2014–2015. He is a contributing author to the Fifth Assessment Report of the Intergovernmental Panel on Climate Change, lead author of the European Transdisciplinary Assessment of Climate Engineering (EuTRACE) report, and chair of the Steering Committee of the Climate Engineering Conference (CEC) series. Schäfer studied at the University of Tübingen and at Cornell University and holds a master's in political science, philosophy, and history from the University of Tübingen. He recently submitted his doctoral dissertation in political science at the Freie Universität Berlin.

Dane Scott is director of the Mansfield Ethics and Public Affairs Program and associate professor of ethics in the College of Forestry at the University of Montana. Scott's primary research focuses on ethics and emerging technologies, particularly biotechnology and climate engineering. He has published numerous articles on agricultural biotechnology and climate engineering. He is editor of *Debating Science: Deliberation, Values, and the Common Good* (Prometheus, 2011). Scott was principal investigator on a recently completed National Science Foundation project, "The Ethics of Geoengineering: Investigating the Moral Challenges of Solar Radiation Management."

Stefan Skrimshire is a lecturer in theology and religious studies at the University of Leeds, where his teaching and research combines continental philosophy, political theology, and environmental ethics. He is interested in the theological and philosophical dimensions of political responses to climate change and the notion of the Anthropocene. He is the author of *Politics of Fear, Practices of Hope* (2008) and editor of *Future Ethics: Climate Change and Apocalyptic Imagination* (2010).

Marit Trelstad is professor of constructive and Lutheran theology at Pacific Lutheran University in Tacoma, Washington. Her scholarly work combines feminist, process, and Lutheran theologies and has focused on Christology, theological anthropology, the doctrine of God, and science and religion (including economics, geoengineering, and ecology). As a contributor and editor, she published *Cross Examinations: Readings on the Meaning of the Cross Today* (Fortress, 2006) and contributed to *Transformative Lutheran Theologies* (Fortress, 2010) and *Lutherrenaissance: Past and Present* (Vandenhoeck & Ruprecht, 2015) along with several other journals and books.

Index

www.ingramcontent.com/pod-product-compliance
Ingram Content Group UK Ltd.
Pitfield, Milton Keynes, MK11 3LW, UK
UKHW041125071026
15587UKWH00009B/205

9 781498 523585